Barriers into Bridges.

The Inclusive Use of Information and Communication Technology for Churches in America, Britain, and Canada

by John Jay Frank, Ph.D.
July 26, 2015
Updated 2016
Published by Minstrel Missions LLC

minstrelmissions@gmail.com
www.minstrelmissions.com

But Now Bring Me A Minstrel . . .
(2 Kings 3:15, KJV) or
(Ezekiel 33:32 KJV)

Turning Barriers Into Bridges:
The Inclusive Use of Information and
Communication Technology for Churches in
America, Britain, and Canada

ISBN: 978-1-887835-22-0
Published by Minstrel Missions LLC
minstrelmissions@gmail.com
www.minstrelmissions.com

Available in a paperback and in a Kindle format.

Keywords and book categories:
1. Ministry and Evangelization, 2. Technology and Church Growth, 3. Missiology, missions, and Information and Communication Technology (ICT), 4. Disability and the Bible, 5. Inclusion or disability discrimination in church, 6. Bible study, 7. Christian nonfiction.

Preface:

My thanks go out to those who read and commented on this book. Some, especially those with the hidden impairments I write about, a vision or hearing impairment, or limited reading ability, were ecstatic at this expression of their experiences, concerns, and the real and reasonable solutions available to churches. Some suggested additional concerns they wanted me to include, but those, while also very important, were not specifically related to the focus of this book, which is how we use computers and information and communication technology.

Some readers who had not experienced the barriers we create in church did not know how difficult it can be to discuss and initiate even easy-to-do changes. They were surprised by the many opportunities now available to us to include those we may have excluded. Please consider this; as a result of this book, one volunteer was asked why she made the print on the weekly church bulletin so small, since some people could not see well enough to read it. She replied, "I never even thought of that." Perhaps now there is a chance and a hope for change.

This book is dedicated to those who presently are able, and to those who are being disabled, and especially to those who work to prepare a gathering place for all of God's people. – John Jay Frank

> You are the light of the world.
> A city set on a hill cannot be hidden.
> (Matthew 5:14)

My Gracious Master and my God,
Assist me to proclaim,
To spread through all the earth abroad
The honors of Thy name.

Jesus the name that charms our fears;
That bids our sorrows cease,
'Tis music in the sinners ear,
'Tis life, and health, and peace.

He breaks the power of canceled sin,
He sets the prisoners free;
His blood avails for all our sin,
His blood availed for me.

He speaks and listening to His voice
New life the dead receive,
The mournful, broken hearts rejoice,
The humble poor believe.

Hear Him ye deaf, His praise ye dumb,
Your loosened tongues employ;
Ye blind behold your Savior come,
And leap ye lame for joy.

O for a thousand tongues to sing my Great
Redeemer's praise, by Charles Wesley, 1738,
Public domain.

Contents

Contents

The demographic data on vision and hearing impairment
and literacy teach us more of what it means to be human.

Here are some steps and standards for computer print,
screen projection, an LCD, the Internet, and sound systems.

Note from the author: The look of a book may depend on where it is printed. The colors on this book's cover attempt to demonstrate the contrast mentioned in chapter 10, but colors for ink printing can vary and cannot accurately represent computer projection. Different programs and printers "make it fit" by shrinking text, or line spacing, or margins, or by kerning, which resizes letter and word spacing, This may look good, except to people who do not see well. Fonts also vary. The main text of this book is 12 point Arial font which is 1/6 of an inch (about .35 mm) from the top of the tallest capital letters to the bottom of the lowest letters (j, g, p, q, or y). Examples of font sizes are presented on page 92, but these may vary because different versions of the same fonts exist and some are compressed or condensed. For one printing the 12 point font became 11.68, 14 point became 13.6, 16 became 14.98 and 18 was 17.51. This book offers specific steps and precise standards, but it does not give instruction for using programs. An Internet search might help with that. A large print edition (18 point font) is available. This book is also in a Kindle format which is a somewhat accessible eBook with audio capability.

1. Introduction

For better and for worse, technology alters the way we share our messages in church. I can remember after youth group taking the typewritten bulletin to a local print shop to be professionally printed. Back then, it was difficult or impossible for some people to read the small print in a Bible, a hymnal, or other materials in church. Later, churches bought photocopy machines, overhead projectors, computers, and printers. Then along came computer projectors, LCDs, the Internet, cell phones, digital cameras, and homemade DVDs. Today, these are routinely used in church and outreach ministries by people who have some knowledge of how to use computers.

I recently shared with an optometrist some of the ways information and communication technology (ICT) is used in churches which make participation impossible for people with low vision. She replied incredulously, "But they're a church, how can they do that! Don't they know better?" Things have changed since we only had small print Bibles, hymnals, and typewriters. Accessible materials can now easily be bought or created, but are still not available in many churches.

Imagine being invited to a church for a service, or a youth group, or a special meeting to watch an introduction-to-Christianity DVD. Along with warm greetings you receive a bulletin or other material with print too small to read. You sit facing a large screen with small words, poor color contrast, and distracting movement. You may try to listen to singing and preaching, or teaching, which is unclear because of distracting background noise. Books around you, a Bible, a hymnal, and other song books, like the text on the handouts or video are all in small print, Some may be incorrectly labeled large print, even though standard large print books exist, are easy to find, and are not expensive. English is your native language. Anywhere else, you and millions of people with hidden, unobvious impairments such as, not being able to see, hear, or read well, expect, or at least hope for reasonable accommodations, except in many churches where they could easily be provided. Even some house churches and home groups misuse technology.

We might call this evangelization and think people who decide not to return after their first visits are just resistant to the Gospel or maybe they did not like our style. This could also be called disability discrimination, but many in the church do not know what that means. After all, we treat everybody the same. It may be that we have "some" knowledge of what it means to be human and how to use computers, but not enough.

* * * * * * * *

We may be using our sound systems inaccessibly too. The story is told of a man who heard a TV announcer describe every detail of a sporting event, but did not hear his wife call him for dinner. This is not inattentiveness. Men typically hear a low-pitched voice better than a high-pitched one. Consider this; the praise band has a female soprano and a male tenor with a high-pitched voice leading worship. During the service someone indicates they cannot hear the words so the volume is raised. Now everyone is suffering with the painfully loud volume that instead of helping, harms our ears. We lament that more men are not in church, or we think older folks dislike contemporary music, and wonder why there is a decline in church attendance. All the while, a synthesizer or guitar capo can lower the pitch, and computers can transpose and print the music. Instead, we raise the volume and buy more powerful sound equipment.

Introduction

Computers, sound systems, smart phones, the Internet, and digital cameras are all entrenched in churches today and for good reasons. Yet, even before those were popular, laws were enacted requiring access for people with impairments whenever it was reasonably possible. We understand how stairs are a barrier to people who use wheelchairs or walkers, but an accessible entranceway, elevator, or lavatory are just the beginning. Barriers to seeing, hearing, or reading, which are less obvious, block the life-altering messages we share in church. We learned to evangelize in cities where most people live, but we may not be mindful of the people who live in the land of disability. These are people who we are, perhaps inadvertently, excluding from church by our habitual misuse of ICT.

* * * * * * * *

I attended a large city high school where only two thirds of the 1,200 students in my class graduated. We all might wonder how our classmates fared. I managed to get through, even with low vision, because I was and still am an avid reader as are many who eagerly wait for the latest books to come out in a large print, audio, Braille, eBook, or Online version. After high school I went to a secular college, then to a Bible college and later pursued further graduate studies.

One high school friend of mine who could not read well was a gifted artist. He attended art school and is a commercial artist. Another friend who did not read well went to a trade school and earned a good living. Another friend who was not good in reading or math owns his own successful business. All three graduated in the lower half of our class. In many churches they would not feel welcome because they would find it difficult to read the Bibles, hymnals, or other song books, or projected songs, outlines, announcements, bulletins or paper handouts.

Turning Barriers Into Bridges

Many people have difficulty *reading a typical small print Bible and find it difficult to participate in church. Not being able to see, hear, or read well are "hidden" impairments. People learn to keep it a secret, to get along without saying anything about the choices others make which exclude them, such as when a church presents inaccessible, print or projected text, or sound. People learn to "get by," say nothing, or just not attend.

* * * * * * * *

*Poor reading is not only caused by a learning disability or by the lack of instruction. By the 5th grade, essential reading skills such as, tracking, scanning, and fluency may be adversely affected by the smaller fonts and increased concentration of letters, words, and lines on a page. This makes it easy to lose your place. These are not the only reasons for a limited reading ability, but they play a part, a part which we have some control over. Song lyrics, bulletins, scriptures, outlines, or messages do not have to be projected or printed in small font (text or letters).

* * * * * * * *

Our technology may seem new, but this issue is not new. Seventy-five years ago, C. S. Lewis, in The Screwtape Letters (#2), revealed the Senior Tempter's glee that "One of our great allies at present is the Church itself." It greets those who enter with a liturgy they cannot understand and religious lyrics in very small print. Today we can use our information and communication technology for clarifying instead of obstructing our messages.

Fifty years ago we learned, "the medium is the message." Inclusive use of technology in churches sends the message that God is accessible to those of us who are able to see, hear, and read well and is also accessible to our neighbors who are less able. Marshall McLuhan also noted, "we shape our tools and then our tools shape us." If we allow the Bible to shape us first, we will be trustworthy, living examples of the Gospel.

Introduction

Consider what Christian workers go through. Pastors, evangelists, missionaries, church secretaries, school, college, and seminary teachers, all are immersed in a very competitive culture where efficient, new technology is highly valued. During an ordinary day, they may rarely or never relate to anyone who cannot see, hear, or read well. They may simplify sermons or lectures, or give one-on-one help, or tutor some, or pray for people, or engage in charitable activities. In their daily work context, however, to habitually produce large print or make other modification for youth or adults is rare. Small print is normal to them and to those they typically interact with. They may learn they are required to be inclusive by law, or by policy, or by a professional code of ethics they signed, but they never need to do it, so they never think about it, or learn how to do it.

New technology for presenting lessons or sermons is helpful (Anderson, 2013), but accessibility is essential. Many of us still distribute small print and project hard to see outlines, scriptures, and song lyrics. Industry is tossing must-have, new programs and devices at us. Attending to one more thing may feel overwhelming and being told we are making mistakes or doing something wrong is unwelcome. We prefer suggestions of better ways to do things, rather than hearing about God's commandments or the secular legal requirements we violate.

Some pastors and teachers have wider experience and produce large print versions of handouts (16 to 36 point font) and larger projected material. Some churches provide large print song lyrics in folders or three-ring binders. One Bible school teacher I know shaved off his mustache so a student who cannot hear, but who can read lips was able to participate in class. Many however, rarely consider access for people with sensory impairments or mistakenly think nothing can be done in church or on a mission field for those who cannot see, hear, or read well, or they may think something will be done by other volunteers or specialists after a request for help is made.

Of course, not every church owns or misuses ICT. In addition, some systems are simply too small for an auditorium, and for some churches, priorities and resources must be weighed because a hearing assistance system can cost as much as a computer system. Also, in some churches pastors do all the work. They manage and maintain the facility. They pray, study, preach, teach, plan for growth, including learning new technology, and care for the members. This can be a huge workload, although it is not unlike entrepreneurs in other fields. Even so, it is possible to learn to use technology in inclusive ways all the time and to begin doing it at a reasonable pace.

* * * * * * * *

This is not an advertisement for new and better computer equipment. This book is based on Biblical reasons for being inclusive and on hidden human variations which effect many of us. Several years ago I declined a request from the editor of an international encyclopedia on disability to contribute an article on technology used by people with low vision. This book is not as comprehensive as that encyclopedia and neither is it about special disability technology. Rather, many of the chapters are short lists with only brief explanations and my concern, now as then, is with the technology most churches already own and use which, if it were used correctly, could include more of us.

* * * * * * * *

The vast majority of people who have a vision impairment do see, they just do not see well even with glasses or contacts. People with only a basic literacy level can read somewhat. Most people who have a hearing impairment do hear, but not well.

Introduction

How we use ICT affects everyone, but people with these or other impairments may be blocked from participating fully or at all.

The demographic data on disability for America, Britain, Canada, and the world are given in chapter nine. By way of introduction, consider the impact of our turning communication barriers into bridges. Church attendance estimates (2004-13) indicate, 39% of the adults in America, 12% in Britain, and 20% in Canada, say they attend church weekly. What if that same percentage of people with impairments could attend church?

If attendance were distributed equally, and 39% of the people who cannot see well in America were welcomed in churches by the miracle of accessible technology, perhaps 23 people who cannot see well could participate in each one of the 350,000 churches in America. If by the same miracle, 39% of those who do not read well, and 39% of those who cannot hear well were also included, each church in America could include 112 of those who have a vision, hearing, or reading impairment.

Adding 39% of those who have difficulty seeing, hearing, or reading is only a rough estimate of attendance. Some people may have more than one of these three limitations or may have other conditions affected by ICT. We cannot assume 39% would attend an accessible church. We can, however, say this adds up to almost forty million reasons to use our technology in more accessible ways even though it may not reach everyone.

if all churches became accessible In the United Kingdom and 12% of those who have a vision impairment attended, each of Britain's 50,000 churches might impact 5 people who do not see well. If in addition, 12% who do not hear well, and 12% who do not read well were included, each church might welcome 41 people. This suggests over two million reasons to be accessible.

Turning Barriers Into Bridges

If all of Canada's 30,000 churches were accessible and 20% of those who have a vision impairment attended church, each church might impact 6 people who do not see well. If in addition, 20% of those who do not hear well and 20% of those who do not read well attended church, each church in Canada could be welcoming to 172 people who might otherwise be kept out. Again, there is overlap in the data. Rather than 172 people per church, this number is better described as over 5 million reasons for churches in Canada to use ICT in accessible ways.

People usually do not reveal hidden impairments. The goal is that we cultivate the Holy Spirit fruit of self-control when using technology so as to stop harming people by unnecessarily excluding them from church. To do this, we need to know both the "why" and the "how" of using the technology we already own, in inclusive ways that have been required for more than 25 years.

Inclusive use of ICT is not just a list of laws, demographics, and techniques. Neglecting to explain why we do this is like celebrating Christmas or Easter without mentioning the birth, death, and resurrection of Jesus Christ. The powerlessness and the scandal of the cross are the power of God (1 Cor. 1:23-28; Gal. 5:11). Both how and why we use technology reflect the Gospel. The Word was made flesh—God became accessible to us. Christ was crucified for us—God was disabled, then died, was buried, and rose again the third day. There Is no resurrection without the cross and no crucifixion without disability.

Barriers in church begin by ignoring disability in scripture. Therefore, the next two chapters offer a disability perspective from the Bible. The two chapters after give definitions and laws prohibiting disability discrimination. Disability is not the result of a person's impairment alone. Disability arises if we fail to provide Biblical, reasonable, legally permitted and required access.

Introduction

The following two chapters look at how a church's culture can make it difficult to change, even though restricting the way technology is used is common and any style of church can become accessible. The next chapter mentions some common, often repeated reasons given for inaccessible use of. ICT. Chapter nine presents demographic information on disability.

Chapter ten and eleven offer more than 80 standards for how and why to use computers to create accessible print, screen projection, a Liquid Crystal Display (LCD) and for the inclusive use of sound systems and the Internet, e-mail, and social media. The focus is on that which is easy to do for little or no cost. Something easily done once can be arduous after 500 or 1,000 times, so a little at a time may be the way to begin.

Finally, the reference section contains more than 70 citations. There are more than 140 scripture references in the book, Some of my experience and training that went into writing this book are mentioned in the last section, "About the Author," along with an e-mail address for comments and questions, which I very much welcome and will try to respond to.

* * * * * * * *

A limited, one-size-fits-all model for setting up common spaces and activities is an extreme that calls for the least effort to satisfy a two-thirds majority. As a result, many people miss the worship, Bible reading, teaching and preaching, and other activities. They are not staying home from church by choice. We are actually blocking people's participation in church. Let us instead create access for those who have sensory impairments or limited reading ability all the time. Reaching more people, with computers and sound systems, to help them reach out to God does not require great effort or an extreme of trying to satisfy everyone's unique needs. There is a balance.

Proverbs 3:27 offers a wiser choice, which is, do what is in your power to do. We can easily and inexpensively use information and communication technology to bridge the gap between the Gospel message God has entrusted us with and the millions of people who cannot see, hear, or read well. A better approach is to aim for 95% of the population to begin with. Some will still be left out, so individual help will still be needed when it is available, but the inclusive use of technology by (1) removing barriers, (2) applying universal design, and by (3) providing reasonable accommodation, called adjustments in Britain, is not hard, or a compromise, or extremism. It is wisdom.

A good way to begin is with prayer. To borrow from the Apostle Paul in Ephesians 3:14-21, "Father God, I pray you call all who you desire, to read this book, to understand it and to act on it and that you would accomplish more than we can imagine through it for your glory. Amen and Amen."

A child picked a Trillium for his teacher in an area where picking flowers is illegal. She thanked him for his kindness, but asked the youngster not to do it again since it was illegal to pick them. We can thank those who helped us use computer technology and gently teach them why it was wrong and point to better ways, being mindful that some habits can be hard to change.

****

"With all humility and gentleness, with patience, . . . making every effort to maintain the unity of the Spirit in the bond of peace" (Eph. 4:2-3).

10

2. A Biblical Perspective

Focusing on people with disabilities found in the Bible and not treating them as if they were metaphors or figurative examples are first steps to including them in church. We may have traditionally classified them as "the marginalized," who receive our prayers and charity, or consider them examples of "the weak," which could include everybody's limits. Instead, we can consider at face value, a disability point of view in the Bible and learn from some obvious, but perhaps overlooked details.

(1)   John the Baptist asked Jesus if he were the Christ. Jesus did not say yea or nay. He said, look at how I **relate** to people who do not see, hear, or walk well, or who are lepers, poor, or dead and decide yourself (Matt. 11:3-5). God heals people from afar by his word and his will. He also came in person and spoke to, listened to, and touched people with severe impairments. He helped people see and hear and understand the Word of God. This is the Christ.

This passage shows that the relationships Jesus had with individuals who had a severe, long term impairment are defining characteristics of the Messiah. From a majority, able-bodied point of view, the focus tends to be on fulfilled prophecy, compassion, and power to heal, which misses the perspective of the people he helped and the practical, personal meaning such as, "I was blind, now I see" (Jn. 9:25). We easily overlook this if we have no relationship with anyone who has a long term impairment. Prophecy, compassion, and power to heal are important concerns, but the majority point of reference is not the only one for interpreting the text and context of scripture. A greater empathy with all people is possible, just by how we preach and teach about disability found in the Bible.

11

(2)   Consider an alternate view of Luke 10: 25-37. To "incapacitate" means to deprive of ability–to disable. In many Bibles the Parable of the Incapacitated Traveler is labeled the Parable of the Good Samaritan and often used to discourage chauvinism, or racial and ethnic prejudice, and encourage doing good deeds. This may be because, like the lawyer in this passage, many are seeking to justify themselves. Good deeds are definitely encouraged in scripture, so that interpretation is understandable; "we are . . . created in Christ Jesus for good works" (Eph. 2:8-10). Good works are an evidence of salvation; "So faith by itself, if it has no works, is dead" (James 2:14-17). A reasonable interpretation of this parable however, requires understanding its context beyond racial and ethnic prejudices, to clarify its purpose and to see a disability perspective.

This parable helps demonstrate by the title it is typically given the tendency to ignore a person with a disability in the Bible in favor of other characters we can more easily relate to. Then, a disability perspective here shows an accepting attitude due to being incapacitated and relying on a hard-to-believe-where-its-coming-from help, which is a faith worth emulating. Further, this parable goes beyond good deeds and shows the need to self-evaluate and confess to whom we are, or are not being a neighbor. Then, it shows the need to demolish man-made barriers to God's love, which Jesus did, and which is required for eternal life and for becoming an accessible church.

Loving our neighbor is God's command, but the purpose of the law is to be "our schoolmaster to bring us unto Christ that we might be justified by faith" (Gal. 3:24). The lawyer's first and main question concerns and always does concern eternal life. His second question, "who is my neighbor?" is a barrier to eternal life prompting a rebuke from the Lord with this parable.

Edersheim (1896) noted that in this parable, Jesus turned the lawyer's question "Who is my neighbor?" into "To whom am I being a neighbor?" He has us ask, "what am I doing?" before "what should I do?" that is, self-evaluation and conviction before action. If instead of trying to trap Jesus, the lawyer admitted he could not fulfill the law and needed, like the incapacitated traveler, a hard-to-believe-where-its-coming-from Savior, even one from Nazareth, his first question, "what must I do to inherit eternal life," could be answered. The parable's purpose is to rebuke the all-too-common Rabbinical wrangling over to whom one must be a neighbor.

Kistemaker (1986) also sees that Christ in this parable is revealing God's intent and meaning to "love your neighbor as yourself," by demolishing man-made barriers that obstruct God's law, in this case, the Judaistic rules of whom a Jew could or could not help. Today, misuses of information and communication technology are barriers that block the Gospel and need to be changed. We need to not only do good deeds, we need to recognize and demolish technology barriers.

In addition, there is a crucial perspective in the parable of an Incapacitated Traveler, in need of help and receiving a hard-to-believe-where-its-coming-from help, We are unable to save ourselves, but are often unwilling to admit it. Instead we look for a good deed to do so we can be righteous by our own will and strength. We cannot earn salvation; ". . .we hold that one is justified by faith apart from works of the law," (Ro. 3:28) and ". . .we know that a person is not justified by works of the law but through faith in Jesus Christ" (Gal 2:16). The parable's salvation-by-grace perspective may be lost if the focus, like the extra-Biblical title it is often given, is the Samaritan's good deeds instead of Jesus' rebuke of making barriers.

13

In most churches, some people do good deeds like the Samaritan and some do not, like the priest and the Levite. Emulating the Samaritan's compassion is commendable, but Christ specifically focused on people with disabilities and he gave us a new law. The old law, to love your neighbor as yourself, impossible as that may be, still is easier and not as sacrificial as Jesus' command to love one another as he loved us (John 13:34, 15:12). We need to examine ourselves and the barriers we build and hold and admit our need for grace to love others sacrificially, as Jesus loves, not merely as we love ourselves or those who are like us.

(3)   Another barrier to becoming an accessible church is to reject the application of scriptures. Textbooks that deny the Bible applies today undermine the moral fabric of our nations. In their book on ethics, Feinberg and Feinberg (1993) assert that the Bible's command to love our neighbor is general enough to apply to us, but putting up rails on a porch (Deut. 22:8) is specific to a particular culture and not relevant today. When everything is reduced to one word, "love," we miss the outworking of love and fail to learn how love works. Evidently, those authors never applied for a permit to build a porch.

Rails with precise specifications are still required today because "love does no wrong to a neighbor" (Romans 13:10). The general principle to build our private and our shared environment so as to avoid harming our neighbor, as well as the many other examples of the application of this principle found in the Bible, remain in force today as moral guides and as secular legal requirements. For many people these are just common sense, but ignoring this is a crime, even if denying its Biblical roots and applicability in a book on ethics published by an evangelical publisher, is not.

A Biblical Perspective

Laws legislate morality. Government and laws are gifts given to us by God for the protection of us all. Long before our secular laws, the Bible declared our responsibility for what we build. Today it is law and common sense to put up safety rails on a porch (Deut. 22:8) or not to dig holes and leave them unguarded (Ex. 21:33-34) and not to lead people astray and not to leave things where people who do not see them might trip over them (Deut. 27:18; Lev. 19:14). Projecting or printing a message clearly for people with low vision is a specific application of these God given laws and the general principle not to harm others. To deny these laws ignores God's opposition to unjust discrimination (Lev. 19:15; Pro. 16:8; Acts 6:1; Ja. 2:1-10). O you who love the LORD, hate evil! (Ps. 97:10).

(4) These next examples illustrate how observing the perspective of people in the Bible with disabilities can help us avoid creating barriers in the first place. Many places, not only churches, have a policy of being prepared to help, but to wait until someone asks for accommodation or adjustments before making anything available. Being willing to provide access after a request is made may sound like adequate consideration, but it often does not work. People are reluctant to ask for help and this is not new. We think the squeaky wheel gets the grease, but all four Gospels show that people were antagonistic toward those who did not see well who requested help.

Two men who were blind shouted at Jesus for help and the crowd following Jesus rebuked them and told them to be quiet (Matt. 20:31). When people who were blind came to Jesus in the temple the leaders were indignant (Matt. 21:15). When Bartimaeus, who was blind, cried out to Jesus for help, many followers of Jesus rebuked him and told him to be quiet (Mk. 10:48). When a man cried out for help seeing, those who

15

led the way rebuked him and told him to be quiet (Luke 18:39). Religious leaders insulted the man born blind, whom Jesus healed, and they threw him out of the synagogue (John 9:28 & 34). This was pre-crucifixion, pre-resurrection, and pre-Pentecost. Is this still a valid disability perspective today? Are people with impaired vision or with any severe impairment still being discouraged from seeking help by the responses and reactions of Christian followers and leaders? Is the response to a request for help today still, "Be silent, sit down," or worse?

A hostile response to asking for help is still experienced. Potok (2002) noted that people with disabilities are expected to act in a manner that appears docile, unprovocative, and undemanding, but requesting legally required help violates that social obligation. In a study of people with disabilities other than a vision impairment, Albrecht and Devlieger (1999) found many said requests for access were seen as making waves that led to their being ostracized, which was even worse than not having access. They would not ask for accommodation because they discovered that asking for help lowers their quality of life even more than not being able to participate.

This avoidance of help-seeking is not due to feeling embarrassment or to a dislike of being treated like a child or a charity case. People in churches may excel in charity, but also obscure or even object to the difference between charity and justice. The avoidance is due to the active hostility expressed toward those who ask for access to what everyone else has. It is often largely due to being told by words and by deeds that accommodations or adjustments, which are integral to one's life and to the lives of many others and which are required by law in most places, are considered odd, or extra, or are being done with great effort, or as an inconvenient favor.

16

Another reason not to ask for help is that asking often does not help (Frank, 2000, Johnson and Frank, 2004; Frank and Bellini, 2005). Asking is futile almost half the time and can be frustrating, difficult, and even dangerous. Frank (2006) found 43% of 311 requests for reasonable accommodation by 151 people with limited vision in America were not fulfilled. The incidence of unfulfilled requests ranged from, 60% of requests for accommodation not fulfilled by government entities, 48% not fulfilled by businesses such as stores, banks, or medical facilities, 37% were not fulfilled when made to employers, and 33% of requests for access were not fulfilled when made to schools. Further, the help that is available may be inconsistent and unreliable when only volunteers are part of the process. Accommodation may be offered, but not really exist, or be hidden, or be faulty and not function well. In addition, many people do not know what to ask for, or who to ask, or how to ask. For many, the burdens are not worth the limited gain. It may be better not to ask and not to participate in any activity where access is not obvious and readily available–including church. Just as safe and unsafe areas become known; so too, places that are accessible or inaccessible become known.

A disability perspective in the Bible includes the view by people with impairments who face disability and our own self-examination. Such a perspective enriches our view of Christ (Christology), salvation (Soteriology), Bible Ethics, and Church outreach (Missiology). An appropriate ending to this partial look at a disability perspective in the Bible is what the Word of God says about the end of time, the final judgment (Eschatology). The end comes after the Gospel is preached to all (Matt. 24:14). That "all" certainly includes people with impairments (Luke. 14:13). If we exclude them now, when comes the end? Perhaps God is waiting for us to get this right.

17

(5)   In what may be the first systematic theology on disability, Yong (2010) notes that Jesus' description of the final division between the sheep and the goats indicates the values by which the world will be judged (Matt 25:31-46). How we treat the least members of Jesus' family determines our eternal destination. Matthew 7:1 and Romans 2:1 note it is not up to us to judge each other. There is one judge of us all who disciplines us for our own good and yet, if we judge ourselves we need not be judged along with the world (1 Cor. 11:31). Considering God's view can lead to self-examination which can lead to conviction, confession, repentance, and forgiveness, which can lead to reconciliation, renewal, and revival.

This brief look at a Biblical disability perspective shows:
1) The Christ we follow is identified by his helping people who did not see or hear well or understand the Word of God.
2) He demolished man-made barriers to our loving our neighbors, especially barriers harming those who are disabled.
3) We are instructed in the Bible to insure that we do not harm, or allow harm to others by what we build or create.
4) Leaders and followers of Christ resisted requests for help from people with impairments, which teaches them not to ask.
5) How we treat the least of Jesus' family will be how we are judged by him.

Therefore, the silence about our misuse of technology must be broken. We could enable instead of disable by printing and projecting larger text, with better color contrast, without distracting designs, pictures, or movement. We could use larger print signs indoors as well as outside and add some descriptive and audio captioning to video. It is possible to build accessible Web sites and install doors that a person who is frail or uses a wheelchair or walker could open.

Rather than evoking mere physical excitement with the power of electricity, we could lower the pitch and the volume, and the electronic effects of our sound systems so people just home from the hospital after another round of chemotherapy, a heart attack, or a stroke could endure our worship and attend church again. We could install a hearing assistance system to our sound systems for those with hearing loss and advocate for Sunday bus service. With these in place we would not need to rely only on volunteers to provide access to church.

Allowing freedom only up to the point before it harms another is a moral value of many cultures. The Royal Law of Love, "Love does no wrong to a neighbor" (Rom. 13:10), includes not misleading or tripping our neighbors with our information and communication technology. We can easily be accessible before anyone asks for help. We can be pro-active rather than reactive at home and on the mission field. We can teach the many Biblical reasons for this in Christian schools, churches, colleges, and seminaries and through Christian media outlets so we may learn to include those whom Jesus told us to bring to our feasts of worship and the Word, "the poor, the crippled, the lame, the blind" (Luke 14:13).

If we end disability discrimination in our churches, we will be lights in the world. The next chapter presents an applied model of accessibility that was planned before the foundation of the world. It was not left up to those who needed it.

Cursed be anyone who misleads a blind person on the road. (Deut. 27:18).

You shall not revile the deaf or put a stumbling block before the blind; (Lev. 19:14).

The hermeneutic style of this book begins with St. Augustine's prerequisites of charity towards God and neighbor (1 Tim. 1:5,), and not claiming exclusive knowledge (2 Peter 1:20-21).

* * * * * * * *

Reading begins with understanding words in their usual or most basic sense. Except where a text is obviously allegorical, poetic, or figurative, or contexts, and related passages, or other truths indicate otherwise, it should be taken literally. First observe, then interpret, then evaluate, and then apply. Face value, or prima facie evidence, is sufficient to establish a fact or to raise a presumption of fact until refuted by other facts.

The seeds of the simple Gospel sprout into a relationship with God, leading to changes in our relationship with others and with our world, and with our use of technology – if the seeds grow.

God's response to the Apostle Paul's three requests for his own healing was, "my grace is sufficient for you, for my power is made perfect in weakness" (2 Cor. 12:9). God's grace and power can still be demonstrated through the church. God chose what is foolish in the world to shame the wise; God chose what is weak in the world to shame the strong (1 Cor. 1:27).

3. Apps and Analogies

In the previous chapter, scriptures were highlighted from a perspective of people with a severe impairment. Hundreds of scriptures mention people with impairments. Many more can apply to the topic when viewed as analogies, that is, they contain similarities suggesting a comparison between a scripture's precise focus and certain aspects of disability. Applying analogies can be difficult. Application is referred to as going from preaching to meddling. It may help if we first compare two kinds of fruit of the Spirit.

Over the past two or three decades a chasm was created by our misuse of technology. A bridge across that divide must be built from both sides. Those who have been disabled and excluded by the churches' ICT must cultivate the fruit of love, such as, turning the other cheek and continuing to engage and be involved with the local church. The disabling church must embrace a greater, but still Biblical range of human variation and also learn new ways of working, which are made possible by cultivating the spiritual fruit of self-control (Gal. 5:22; 2 Peter 1:6). The motivation and aim of all we do should be love (1 Tim. 1:5; 1 Jn. 4:7-8), but the fruit, or steps needed along the way, are for us to exercise self-control and to publicly let it be known that we aim to be accessible.

To understand the differences between good deeds motivated by love and kindness and practicing responsible self-control, consider that when giving someone a ride in a car we see the person helped. In contrast, when we fulfill the law by driving safely, we usually do not see the people we did not hurt or the accidents we avoided. The difference is also highlighted by Paul's concern that he not be disqualified (1 Cor. 9:27).

21

Turning Barriers Into Bridges

Suppose Christians, in their home country or abroad, on their way to love motivated ministries, drove at twice the speed limit and ran people over. They would be disqualified. Misuse of technology matters. Ophthalmologists perform surgeries in mission hospitals around the world. Members of Lions clubs collect old glasses and with the help of optometrists they too help people see. Old hearing aids can also be recycled. In some churches, volunteers teach reading, English, or computer word processing, or create large print or Braille Bibles to give away. These uses of technology demonstrate the fruit of love, but they would all be unacceptable and disqualified without self-control in how the technology is used.

Paying an electric bill, obeying traffic laws, or using ICT inclusively instead of exclusively are a few examples of specific behaviors needed when using any technology. These are not good works or acts of charity. They may include the fruit of love, but we usually refer to them as personal responsibility, discipline, and self-control even if those terms are not popular.

* * * * * * * *

The following four analogies are examples of God's love and self-control and his inclusive communication. The greatest ACCESSIBILITY story ever told is found in John 1:14, "And the Word became flesh and lived among us, and we have seen his glory, the glory as of a father's only son, full of grace and truth." and in Eph. 2:17-18, "So he came and proclaimed peace to you who were far off and peace to those who were near. For through him both of us have ACCESS in one Spirit to the Father." We are redeemed because God revealed himself in the ACCESSIBLE form of his Son, "through whom we have obtained ACCESS" (Ro. 5:2). We have ACCESS to God because Christ died for our sins, was buried, and was raised on the third day (1 Cor. 15:3-4). Do we deny it to others?

22

Apps and Analogies

The most astonishing example of **DEMOLISHED BARRIERS** to God's love is noted in Mark 15:35, "And the curtain of the temple was torn in two, from top to bottom," and Eph. 2:13-14, "now in Christ Jesus you who once were far off have been brought near by the blood of Christ. For he is our peace; in his flesh he has made both groups into one and has broken down the dividing wall, that is, the hostility between us." He is still breaking down and demolishing the unnecessary barriers we build, even the walls of technology and our overwhelming busyness with everything else that needs to be done. This is not a call for extra work or to do more good deeds. We need to relearn what we already do, but learn to do it correctly, accessibly. This is also called repentance, which is recognizing and admitting we are doing something wrong, stopping and turning from it, and going in another direction. It allows forgiveness and reconciliation, which is renewal and revival.

The most user-friendly **UNIVERSAL DESIGN** is found in Matt. 11:28 (NKJV), "Come to me all you who labor and are heavy laden and I will give you rest" and John 7:37, "Let anyone who is thirsty come to me," and Rev. 22:17, "And whosoever will, let them take of the water of life freely." All who will come can come and partake, not only those who are able-bodied. Not everyone will come, but those whom the Father calls will come. We are called upon to help prepare the way for them—a way without stumbling stones (Isaiah 62:10).

The most valuable **ACCOMMODATION OR ADJUSTMENT** ever provided to helpless, fallen humanity is described in Eph. 2:8, "For by **GRACE** you have been saved through faith, and this is not your own doing; it is the gift of God." We neither paid for, worked for, nor deserve this precious accommodation that allows us to participate in God's Kingdom and to share the good news with others, "Freely you have received, freely give"

(Matt. 10:8 NKJV). We are called to be imitators of God (Eph. 5:1; I Jn. 2:6) who willingly became accessible to us. Being an accessible church sends a message of how accessible God is and the message of God's grace to us and through us.

* * * * * * * *

"The hearing ear and the seeing eye, the Lord has made them both." (Pro. 20:12). God's opinion of ingratitude is found in the parable of the man who was forgiven a huge debt he could not pay, but who then did not forgive a very small debt which he was owed (Matt. 18:23-35). Three quarters of the population in America (about 225 million people) use glasses or contacts. We show our thankfulness for those technology gifts that help so many of us by using our other technologies in church to include people who cannot see well even with glasses or contacts. When we provide large print or accessible computer projection, which are essential for millions of people, we demonstrate we are thankful for our many gifts.

The telephone was invented as an aid for people with a hearing loss. We can show gratitude for that beneficial gift of technology by using our sound systems in church to include people who do not hear well. Instead of increasing the volume to where it harms people's hearing, we can lower the pitch and add a hearing loop or other hearing assistance system. Human variation and disability impact us all. What will be our grateful response to God's grace to us?

We have habitually set up church using technologies incorrectly every day or week for decades. The problem is not a scarcity of volunteers to do extra work on additional priorities, Whenever it is reasonably possible and does not alter the essential nature or purpose of an activity or space we need to learn to be inclusive to begin with, before we set things up incorrectly. The knowledge is not hidden. It has been ignored or rejected and bad habits can sometimes be hard to change.

* * * * * * * *

A Biblical view of disability, whether precisely found in scripture or inferred by analogy, does not omit advocacy. The Bible goes beyond studies of attitudes or social stigma. The Bible calls for advocacy, which is an action that produces changes in behavior. "Speak out for those who cannot speak, for the rights of all the destitute. Speak out, judge righteously, defend the rights of the poor and needy." (Pro. 31:8-9).

Timid Christianity is uninspiring and lukewarm. Who will display the courage of John the Baptist and preach about the incidence of divorce and its consequences when a family member experiences a disability (Singleton, 2009)? Who today, like Jesus, is overturning the money tables piled high with illegally inaccessible notes? Years ago, the United States Treasury Department lost its court case and was ordered to print accessible currency like many other countries do, but the U.S. Treasury still does not do it (O'Connor, 2013). Our silent voices must become louder than the hawking shouts in the technology marketplace.

* * * * * * * *

If analogies between redemption and including people with impairments in church seem like a stretch, or like reading my own ideas into scripture, consider whether they contradict any other part of the Bible. Since all have sinned and fall short of the glory of God (Ro. 3:23) does any scripture specifically exclude people with impairments from approaching God? Does Lev. 21:16-23 mean people with impairments are not allowed to serve or to approach God? Perhaps God excused them from the arduous labor of Temple worship out of kindness. It could be that the 24/7 shift work of hauling water, wood, ashes, and slaying animals and not their blemishes, would profane His sanctuary. God is not a cruel taskmaster. He allowed them to partake of the most Holy bread and the Holy things. Do we?

Turning Barriers Into Bridges

We can include people who have low vision or hearing as leadership, staff, and volunteers in church and Christian ministry. Think of Sunday school or Bible study teachers and helpers trying to read a small print, computer-generated class list. We can make at least some printed material, including sign-up sheets, in larger text (14-16 point font). Can we install a sound system with a hearing loop in some classrooms?

A large print Bible (see page 83) was essential for me as a pastor and as a chaplain of a nursing home, a college, and a hospital. As a counselor at a Christian rescue mission I reformatted the handbook everyone used into larger font. As a house supervisor at an Arc group home and when helping at a Friendship Club I found larger print would help with such practical materials as the directions posted for a heater, an air conditioner, a hot water unit, and a list of emergency phone numbers. My training as a therapy aid at a State psychiatric center included distributing medications, a critical area where larger fonts help. Making such changes can allow people with and without impairments to grow together in the grace and knowledge of our Lord and Savior Jesus Christ (2 Peter 3:18).

* * * * * * * *

Secular laws prohibiting disability discrimination are described in the next chapter. These pale beside the Great Commandments (Matt. 22:38-39: & John 13:34;15:12) and the Great Commission (Matt. 28:18-20), but they are closely related to both. The Bible says, "it is not in man who walks to direct his steps" (Jer. 10:23) and "Let every person be subject to the governing authorities" (Rom. 13:1-7). "For the Lord's sake accept the authority of every human institution" (1 Peter 2:13-16). Jesus said of a secular law, "if anyone forces you to go one mile, go also the second mile" (Matt 5:41). These laws are part of the context and the environment of our churches.

26

4. Grace and Law

Media disclosure of horrid, segregated conditions led to a public outcry and then to deinstitutionalization. The National Association for Retarded Children changed its name to The National Association of Retarded Citizens, then to The Arc of the United States because people with cognitive impairments rejected the degrading names and demanded change. "Blind Men Marching" gained laws for their own self-determination (Matson, 1990). Parents protesting won IDEA so that their children can attend school. Vietnam veterans in wheelchairs fought for and won the ADA (Chariton, 2000). To paraphrase Edwin Markham, "They used technology to keep us out. Contemporary ambiance, aesthetics they shout. But Love and I had the wit to win. We passed a law that allowed us in."

Discrimination on the basis of disability has been illegal for more than 25 years in the United States under the Americans With Disabilities Act of 1990 (ADA), now the ADA Amendments Act of 2008 (ADAAA, 2008). It has been illegal for more than 20 years under Britain's Disability Discrimination Act (DDA) of 1995, (now part of the Equality Act [EA] of 2010) It has been illegal for more than 30 years under the Canadian Charter of Rights and Freedoms of 1982 and since 2005, with enforcement provisions, under the Accessibility for Ontarians with Disabilities Act (AODA). These laws apply to churches in all three countries, yet mention of the laws and the term "disability discrimination" is rarely or never seen in Christian media. Secular media is equally silent (Johnson, 2003).

Laws against disability discrimination exist because we create much of our environment and produce the things we use. We can regulate the things we build and manufacture. Creating equal access means owning responsibility for the

27

consequences of what we create. The debilitating result of poor vision or hearing, or limited reading ability is not caused by an individual's impairment alone. We create barriers. It is not necessary to only produce small print, project hard to read messages, or misuse a sound system. We could be inclusive to begin with and not wait for individual requests for access.

Definition: Disability occurs in the interaction between features of a person and features of society. Access for people with impairments requires removing, where it is reasonably possible, physical and social barriers. Access is thwarted if churches never learn about the laws and methods to establish a clear and comprehensive end to disability discrimination or if they wait and only apply them one person at a time.

Laws prohibiting disability discrimination are similar. The Bible and secular law often, although not always coincide. The laws involve more than can be covered in this brief book. This is only a slight sketch of the legal context and environment of churches for the purpose of introducing the definitions and laws and how they work. This is not offered as legal advice. Sources for further information are provided for each law.

United States: In enacting the Americans With Disabilities Act (ADA), the United States Congress found that people with disabilities face discrimination barriers that keep them from participating in **all** aspects of society including such critical areas as **communication**. The law's purpose is to eliminate those barriers by providing a comprehensive national mandate with clear, enforceable standards. The ADA prohibits several types of disability discrimination, including; the failure to build in accessible ways for people who have a seeing, hearing, walking, or other impairment, or the failure to remove obstacles when it is reasonably possible to do so; or creating inaccessible activities that could reasonably be made inclusive; or failure to provide reasonable "accommodations" (called "adjustments" in Britain) that allow participation by people with disabilities. Creating access is required both in general and for individuals.

Those definitions of disability discrimination are used in local laws and in other countries as well. "Reasonably" or "reasonable," means that the law does not require altering the essential nature of an activity or space and also that adherence to the law should not be too costly or too difficult (ADA, 2009, U.S. Department of Justice, 2012). About 20% of the population has a severe impairment, so using about 20% of our resources for the purpose of accessibility is an estimate of the terms "not too costly" and "not too difficult."

Everyone in the United States is protected by the ADA, but churches are exempt from much of the law due to limits on Federal government involvement with religious organizations. A church with five or more employees, like any business, may not discriminate against present or potential employees due to their having an impairment. However, if people with an impairment are blocked from participating in church, how could they acquire a desire to work there?

Bickenbach (2000) observed that the laws prohibiting disability discrimination are complicated to utilize and enforce. They may be unknown or not understood. Government relies less on litigation and more on education and mediation to enforce the ADA, however, the complaint process in the U.S. results in people "winning" nothing except a too costly right to sue (Moss, Burris, Ullman, Johnsen, and Swanson, 2001). Then, the court system results in an overwhelming number of losses (O'Brien, 2001). Forced compliance or compensation is rare. The ADA amendments of 2008 provided clarification of the law's intent, but in order to gain that, a compromise was required which further weakened its enforcement. Beyond Paul's admonition that believers should not go to court against believers (1 Cor. 6:1-8), people with impairments have learned to forgo the costly, burdensome, and often ineffectual ADA complaint process (Frank, 2006). Moreover, who wants to participate in a place where you have to force your way in?

State and local governments in America regulate churches now. Access for people who use canes, walkers, or wheelchairs has improved in churches due to local laws which require accessibility be part of any plans before granting building, renovating, or remodeling permits. We can also be thankful that churches are subject to local health, sanitary, and safety laws, such as laws for fire exits, or the number of people allowed in a building, and safe electrical wiring. Nevertheless, after the electrical wires are connected from the outside, what we do with computers and sound systems inside churches in America is largely unregulated unlike the laws for nonreligious organizations and unlike in countries that do not emphasize the separation of church and secular government.

Britain: The Disability Discrimination Act (DDA) of 1995 requires "reasonable adjustments," that is, service providers must provide access to goods, facilities, services, education, and premises. This includes information and communication. Churches in the UK are covered by this law and the later Equality Act of 2010. It is unlawful for service providers to treat people with disabilities less favorably for a reason related to a disability. Service providers may have to make reasonable adjustments in relation to the physical features of their premises in order to overcome physical barriers to access.

30

The Equality Act of 2010 (EA) combined the DDA with other nondiscrimination laws. It also protects persons associated with a disabled person such as caregivers. For more information, see the free, 19 page handbook, Disabled Consumers and the Equality Act 2010, at:
www.dls.org.uk/advice/factsheet/consumer_contract/ goods_services_EA/Goods%20and%20Services%20and%2 0the%20EA.pdf Also see the free, 26 page handbook, Enforcing your Rights as a Disabled Consumer, at:
www.dls.org.uk/advice/factsheet/consumer_contract/enforce_ rights/Enforce_Your_Rights_As_A_Disabled_Consumer.pdf.

Under the DDA the government provided some legal advice and in 2000, the Disability Rights Commission (DRC) received additional power to pursue agreements in lieu of enforcement action. The act was strengthened in 2005. Although the Equality and Human Rights Commission has some enforcement powers in certain situations, its limited resources means it cannot handle many cases. Under the EA of 2010 it is still up to the individual to go to a court or tribunal to lodge a complaint. The process is largely an individual ordeal against a large organization, undertaken by a person who has a severe impairment.

Dickens (2008) notes that having different standards for public and private sectors makes the EA less efficient. Also, its emphasis on resolving individual complaints rather than seeking changes in the discriminatory behaviors of organizations makes it even less effective. The fact that complainants have to try to have their rights enforced in an adversarial tribunal system where procedures are complex and slow, and where most unrepresented applicants are at a disadvantage, calls into question the equality of access to justice. Despite changes designed to streamline the EA, Dickens (2014) is not optimistic about its effectiveness.

Canada: The Accessibility for Ontarians with Disabilities Act (AODA) became law in 2005 with full implementation in 2012 and with the goal of an accessible Ontario by 2025. It is a continuation of several other laws protecting people with impairments, such as the Canadian Charter of Rights and Freedoms of 1982, the Canadian Human Rights Act of 1985, and the Ontario Human Rights Code of 1992. The Ontario Human Rights Code prohibits overt discrimination as well as practices that are discriminatory in their effect.

These laws establish the principle of equality of access for persons with disabilities to goods, services, facilities, and employment. The law requires being accessible, not merely providing access one case at a time. It applies to both public and private organizations, including churches and educational institutions. Almost twenty years ago, I discussed this issue with scientists and government officials at the Ministry of Public Health in Ottawa, Canada. They knew the scope of the issue and devoted attention to how the country should proceed. Laws were proposed based on predictions about the future population. The laws and the predictions have come to pass.

The AODA provides legal mechanisms for individual complaints, enforcement, and penalties. The Accessibility for Ontarians with Disabilities Act Alliance (AODA Alliance, 2015) reports the government is not enforcing the AODA in public or private sectors despite having over 20 million dollars earmarked to enforce it. This is another David versus Goliath story with the twist that David may be unable to see, or hear, or read well.

The AODA requires that **communication** with a person with a disability be in a manner that takes into account a person's disability and that staff be trained on serving and interacting with people who have disabilities. It also requires an accessible feedback process so the public will know both the accessibility status of an organization and of a particular request for access. Guidelines for the AODA are available free, Online in the 77 page .pdf document, "Accessibility Standards for Customer Service Handbook" (2009) (www.mcss.gov.on.ca /documents/en/mcss/accessibility/Tools/AO_EmployerHand book.pdf). Also available at no cost, Online is the 44 page pdf booklet "Information and Communications Standards" (2010) (www.mcss.gov.on.ca/ documents/en/mcss/accessibility /iasr_guidelines/ Part2_IASR_2012.pdf).

The enforcement of the AODA in Canada is likely as difficult and rare as the enforcement of the Equality Act in Britain and the ADA in America. Local city laws may be more effective. Laws prohibiting disability discrimination that are largely unenforced and ineffectual apart from the good will of the citizens are the context and the environment of the church.

We may feel no urgency to use our information and communication technology in accessible ways because we have little or no fear of civil penalties. Or, we can be inspired to reach for our high calling to be the light of the world by being obviously accessible. That choice would be ours to make, if we knew it existed: if the laws and scriptural reasons they exist were preached, taught, and discussed in church, Christian schools, and in Christian and secular media. Instead there is silence or misinformation and opposition. Biblical and secular laws against disability discrimination are too often ignored or unknown, diminished, exaggerated, or rejected.

The United Nations: The UN Convention on the Rights of Persons with Disabilities (CRPD), adopted in 2006, has been ratified by 151 countries and includes similar definitions and many of the same requirements that were mentioned above (see: www.un.org/disabilities/convention/conventionfull.shtml). The UN World Health Organization (WHO) distinguishes between people's impairments and the disability which is caused by their interaction with certain aspects of society. The UN CRPD acknowledges that persons who have disabilities, in all parts of the world, continue to face barriers to their participation as equal members of society. It recognizes the importance of accessibility in all areas, but especially to **information** and **communication** systems.

The signers of the CRPD, which include Canada and Britain, agree to eliminate disability discrimination and barriers to accessibility, including the refusal to provide reasonable accommodation or adjustments, in order to ensure that persons with impairments have equal access. The United States signed the treaty, but has not yet and may not ratify it (see www.ncd.gov/publications/2014/07142014/). Churches in America are exempt from some parts of the ADA, not so they can discriminate on the basis of disability, but to insure that the Federal government does not support religion. Local anti-discrimination laws do apply to churches.

The CRPD requires collection of data concerning the impact of the law on people with disabilities and on the removal of barriers. This can be misleading. I evaluated ADA complaint and mediation reports which found thousands of complaints were "successful." They revealed that real barriers leading to valid complaints exist. Unfortunately, successful mediation does not reveal if barriers were removed or if people with disabilities were helped by filing complaints. That more precise information could easily be collected, but is omitted.

The CRPD mandates equal access to justice, but as mentioned for the ADA, the Equality Act, and for the AODA, utilizing the legal system is often difficult and costly. This is not unusual. Laws protecting anybody, against petty crimes, or even for more serious crime, are often not enforced even if an attempt is made. There are not enough police or judicial resources to address all violations of law within our legal systems. Choices of what to prosecute must be made for every law. When disability discrimination law is ignored, opposed, or has a low priority, equal access for people with disabilities means equally slow, limited, or denied justice.

Secular laws articulate moral principles which are often based on the Bible. Nonetheless, except for very grievous, lucrative, or large demonstration cases, equal access is based mainly on the law-abiding character of a society, that is, its moral fabric – which churches help create. The Church is called to be the light of the world (Matt. 5:14). Even so, opposition to these laws exists in secular realms and in churches. When a church is inaccessible there may be no other church to attend. There is little current data on church use of technology or attendance by people with impairments. It can be unpopular and expensive data to collect. Since most people are loath to file a complaint against a church, our disability discrimination is revealed by expenditures and efforts which, instead of being inclusive, only benefit the able-bodied.

A survey 17 years ago found just under half (49%) of people with a disability in America attended church at least once a month compared to 59% of those without a disability (Kaye, 1998). More recent estimates suggest four fifths (80%) of those experiencing disability themselves or in their family can no longer find a church to attend (Vander Plaats, 2014; Stumbo, 2014). Anyone can easily verify those estimates, as I have, by contacting churches by phone or in person.

A visit to a dozen churches in one area revealed they all used technology. First, I asked during the week, to prepare in advance and then, on Sunday when nothing was obviously available, I asked again from the ushers distributing small print material. Three out of 12 provided large print material (by photocopy) and only when it was requested. I also called the main offices of churches and church denominations to ask if they made provision for people with disabilities. The positive responses directed me to separate special education services for people with special needs, such as, those with Intellectual and Developmental Disabilities (IDD). But the people I spoke with in those churches had never heard about modifying how technology is set up for people with a sensory impairment, or who did not read well and they would not do it.

How we use technology may have contributed to a huge decline in church attendance by people who have sensory impairments and their families–who could become any of us. Disability discrimination in the way a church uses ICT has the effect of excommunication and disfellowship of members of the church without even a trial (Ps. 42:1-5). This negatively impacts everyone. Without a Christian ideal to attract the young, with less justice in church than in the rest of society, an inaccessible church is unattractive to the able as well as to those it disables.

Part of the research for this book was to read Christian books and articles on the topic. There are many that focus on people with IDD or "special needs," but almost none mentioned self-control by the able-bodied and technology and the law. Some stressed attitudes were the greatest barrier, which can be true for some types of obvious impairments. Some tried to create a new theology of disability, or felt that the details of access were already well known. Few mentioned disability discrimination with technology or its demographic scope.

Grace and Law

Some books and articles oppose the laws by suggesting they harm the very people they intended to protect. Bagenstos (2004) exposes that "perverse-results" argument as an attempt to give the appearance of supporting the people who face discrimination, yet opposing the laws protecting them, "for their own good." A "perverse-results" excuse in churches may take the form of opposing inclusive use of ICT in order to reach others. That approach is saying, "we will be unable to share the Bible if we abide by the Bible." It denies the power of the Gospel and seeks to continue discrimination. Such opposition is not surprising. Our technology gives us unprecedented potential to fulfill the Great Commandments and Commission.

These laws are also opposed by charges of, "excessive government intervention in church." Such exaggeration tries to promote fear of the law. Another way the laws are opposed is by referring to the crime of disability discrimination as if it were about manners or hospitality. That spin minimizes the law and diminishes our responsibility. Disability discrimination is a crime and a sin. It is not a question of manners or hospitality. People with impairments may be life-long members of our church who would attend and participate if we made it less painful and more accessible. They are not only strangers to whom we should extend hospitality. Charity, often seen as pity, is not the answer (Shapiro, 1994). Substituting benevolence for justice may only offend people who have lived for more than 25 years in countries where technology barriers are illegal.

There are valid questions regarding the effectiveness of laws against disability discrimination in church or anywhere else. Of greater concern is our churches' silence concerning the law, disability, and the Biblical basis and mandate for the inclusive use of technology. If we speak up and attend to an issue there is often measurable success.

My apologies if I missed or misinterpreted any book, but I believe, at least with ICT, behaviors are a greater barrier than attitudes and are more readily changed. Laws do help keep us from committing mean and thoughtless acts. Every time I read through the Bible I see that people with impairments are valued and included in God's family, church, and kingdom. If those scriptures are taught and preached attitudes will change. There has always been a theology and a grand theory concerning impairments, disability, and inclusion. It is called Christianity. Therefore, the requirement to discuss, to know, and to apply the details of accessibility is always needed by churches.

For the first time in history we have the tools to easily include more people. If we demolish man-made barriers, build with universal design, and provide reasonable accommodations and adjustments we can share a message of, "acceptance by God" in a way that says, "we accept you." We can, in humility count others more significant than ourselves and look not only to our own interests, but also to the interests of others (Phil. 2:3-5).

Soon after this book was announced in Britain, a statement was posted on the Church of England's Website that churches in England are subject to the Disability Discrimination/ Equality Act. In less than a fortnight that notice was replaced by pages of outcry against government intervention in churches. A search of that website revealed their most comprehensive disability access audit made no reference to creating accessible websites or to clear print guidelines. Exaggeration and fear mongering occur in America too, also without mention of how to be inclusive. We must demonstrate that we are citizens of a better country, one whose King and kingdom are accessible.

* * * * * * * *

The next chapter explores how technology is often regulated throughout society and even in churches.

5. Restricting Technology

"You shall receive power when the Holy Spirit has come upon you" (Acts 1:8). We also receive power when we use technology, be it a train, a jet plane, a car, a sound system or a computer. We can be faster, louder, smarter, and more powerful with technology and that is hard to restrain. Restricting how we use all types of technology is normal for all of us, even for those who feel they have a vested interest in the "newest thing," and how it is used. Paul's discussion of justice, self-control, and the coming judgment, was frightening (Acts 24:25). Today, discussing the restraint of technology is rare, but it should not be alarming. Regulating how we use technology is normal in both religious and nonreligious organizations.

It may not be our fault that we learned how to use technology incorrectly, or went along with what volunteers did. Still, it is not written in stone and inviolate. At the same time we reach out with charity, good deeds, or evangelizing or missions, we cannot neglect our responsibility to carefully use our latest tools in well-known, inclusive, and nondisabling ways. Following the status quo disables people with impairments and excludes many. We can learn to build bridges, to not harm and exclude people, and instead to include them, by changing what we have been doing wrong for decades with our information and communication technology. Learning some new tricks and breaking bad habits is not that difficult.

The government repeatedly uses media to remind the public of various restrictions on technology, such as, giving warnings of the stiff fines for not obeying speed limits, not wearing seatbelts, or not having the required number of passengers in an express lane. Motor vehicles were used for decades before safety features were required by law. The first standards enacted fifty years ago were seatbelts, dual braking systems, padded dashboards, and windshield wipers. Sadly, years before safety measures were required, the auto industry knew those features would save lives and sold them one at a time to individuals, but refused to apply them universally, claiming the added cost would be unpopular with the public.

Safety issues were ignored until the auto industry and the public were forced by law to act. Thanks to laws restricting the manufacture and the use of transportation technology, the automobile death toll in America dropped from 5.3 deaths per million miles traveled to 1.1 per million miles (the National Highway Traffic Safety Administration, 2014). How many lives could be saved for eternity if churches fulfilled the intent of laws requiring accessibility for people with impairments?

Auto safety laws are not mentioned here because auto accidents cause disability. The point is our reluctance to restrain how we use technology until forced to by law. Recently, we were unwilling to forgo using cell phones to talk or text messages while driving, so hard-to-enforce laws were enacted to protect us from our dangerous use of that technology.

Many colleges prohibit freshmen or sophomores from having cars on or near campus. High insurance rates for younger drivers inhibit some young people from driving until they are older and (hopefully) wiser. Even movies and concerts begin with requests to shut off personal electronic devices because many of us lack self-control regarding the use of our personal technology. Restricting technology is normal. We may not like external controls, but, most of us agree we need them.

Restricting Technology

We like technology, but when it causes harm, we control it, or fix it, or change how we use it, and provide warnings. Many Americans fiercely defend their right to bear arms, but in most communities it is illegal to discharge a firearm. How we make and use any technology–not just ICT is restricted by law. When a church ignores this, it harms the people Jesus told us to bring to our feasts. One pastor asserted he would not use his church's technology in inclusive ways until he was paid to or forced to do it. Perhaps he was thinking of all the work of trying to do it all at once. Church has never been perfect. Acts 6:1-3 describes how a deaconry was created to correct injustice in the food distribution to widows between the Hebrew and Greek disciples Realizing that the church has a technology abuse problem should not be too shocking or too hard to fix.

We can take care of technology misuse ourselves or we will find out that restraint comes in various ways and from various sources. In some areas, power companies vary their rates for electricity to limit consumption during peak demand which controls how we use electric stoves, washers, and dryers. Recently, after years of publicity on the problem of child abuse in church and after expensive lawsuits, churches had to be forced by their insurance companies to comply with safe child practices for Sunday School classes. If we wait, change will be forced on us to our shame. It will not glorify the name of Christ.

Laws restrict radio and television. The 21st Century Communications and Video Accessibility Act of 2010 requires video descriptive programming on TV. This has been available on tours, in museums, art exhibitions, and for sporting events. Descriptive video and audio captioning can be technically simple or sophisticated, short or longer. It can be a spoken part of an introduction or be inserted during a video clip we

41

download or make for church with our digital cameras. Written and narrated descriptions of key spoken and visual elements of a video benefit more than only the people who have a vision, reading, or hearing impairment. Such universal design can aid everyone's understanding and memory.

History and current experience teach us that concrete obstructions created by new technologies need concrete laws, not a new theology or merely changing attitudes. Barriers must be demolished and new behaviors learned. Despite years of disability civil rights laws in America, Britain, and Canada, some churches are still inaccessible. It is worth noting, the younger generation cannot afford to buy new technology or to build and maintain church buildings. Their parents and grandparents pay for this. Misuse of technology is not a generational clash, it is a crime and a sin that harms people of all ages.

None of us will hear, "well done good and faithful servant" because we learned to use the latest computer program, unless we first consider the least among us in the way we use it. The church has influence for good and bad. Our failure to be a light of the world in this area contributes to the ongoing resistance from technology manufacturers, many of whom also refuse to even use existing, inexpensive ways to make their new products and programs accessible despite the laws requiring them to build in accessibility.

Legal restraint is not yet a viable solution to the disability discrimination committed by churches because enforcing one case at a time in court with $5,000 or $55,000 fines is too difficult. A more easily enforceable approach would be to add smaller $50 and $500 fines against the misuse of ICT that can be easily demonstrated, reported, proven, and adjudicated.

Restricting Technology

That level of policing occurs now for cell phone use while driving and it exists with building and fire codes, even for churches. Disability discrimination can be diminished before it occurs if we had more easily enforced laws to target schools that teach inaccessible Library Science, or Computer Science, or any inaccessible ICT. We need laws to target and restrict technology manufacturers and computer program writers, and Internet sites that supply inaccessible material to churches.

Supplements to existing laws against individual acts of discrimination can change entire industries following the successful examples of the National Highway Safety Act, the Communications and Video Accessibility Act, and the Federal Communication Commission's proposed Open Internet Order, or the Net Neutrality Rule which seeks to protect access to the Internet for people who have impairments. Practical pressures and restrictions are possible and do change behaviors. These exist in some places and more are being created, but it is a slow process. Some leadership from churches would help.

Twenty-six states in America require disability awareness training in public schools and colleges now. More will follow. The content of the training is still a question. Churches may not be mentioned at all and the point of view may be that of the institutions rather than the perspective of a person with a disability. Disability awareness training using a wheelchair, wearing mittens, ear plugs, or scratched goggles smeared with petroleum jelly, or other experiences are counterproductive. This often creates pity or fear instead of awareness.

A disability perspective is earned over time. Learning to live with a long-term or a lifelong functional impairment and learning to avoid or overcome disability discrimination is not the

43

work of a day, a week, or a month. That is why the Bible has much to say of eternal value about justice and disability. Short of having an impairment and facing disability, the way to be more aware of a disability perspective is to talk with people who live with an impairment and want to end disability discrimination.

The question of when restriction or prohibition is needed, or is even possible, affects various technologies. We already make such decisions and act upon them. This is not new, but we still have many moral conflicts over technology to resolve. Who will believe or value our input on other technology issues if we will not even provide a large print bulletin?

Churches, as living examples, could more effectively share a Biblical perspective on such medical technology issues as, what we develop and who we allow to live, or keep alive and for how long, and with what technology. Controversy exists in areas such as, conservation and pollution, the disposal of toxic old computers, poisonous pest control and air, water, and land use. and the impact of technology on our inner- and inter-personal relations. These conflicts are about technology use. The Bible and the church have a point of view worth inputting into ongoing discussions concerning technology.

Our source of moral truth, the Word of God, is too often left out of this conversation. Temple with Ball (2012) point out that, we, those who are able and those disabled, are made in God's image and Christ is restoring us into his image (Genesis 1:26-31 and 1 John. 3:2). Inquiry into ICT's affect on people or society that considers only the able-bodied distorts that image. For the last 30 years, mention of ICT in Christian publications and churches was largely about buying it and how to use it, and contrasting styles of music and worship. As a result, today we value technology in the same pragmatic way secular society values it; for efficiency, utility, and fun and we may feel we cannot do without it or change. By ignoring the harm we did and still do and the laws which forbid using ICT in ways that cause disability, we miss the opportunity for repentance and growth. The church could instead guide itself with the godly resources available – the Bible, and deal with its own culpability and responsibility. Then, it will find itself fulfilling its reasonable service as a many membered body (Rom. 12:1-5). The next chapter explores one historic obstacle which those who love the Church must overcome in order to fulfill that reasonable service.

Writing is a technology. We have the Bible due to the accuracy of the scribes who followed the rules when copying it. We are still called upon to use our tools to clearly share the Word of God with as many people as possible (Matt. 28:18-20).

A pastor tried out his new car on an empty road. Aerial surveillance clocked him at 110 mph (about 190 kph). He received a ticket in the mail, but authorities later found out he was in the country on a work permit and was not a citizen, so he was deported. It took him some time to get back in during which time the church was without its shepherd. **The use of technology is restricted and someone is watching**.

Computers and technology are not the only areas pastors find overwhelming. One Sunday, after I spoke in a church of about 50 people, the pastor came to the pulpit and gently asked for volunteers to help clean the church each week, which only his wife had done for many years.

* * * * * * * *

The Code of Ethics for Congregations and Their Leadership Teams, of the National Association of Evangelicals (NAE, 2015), affirms that we will minimize barriers that would discourage persons with disabilities from full participation.

* * * * * * * *

Is technology god or can we—will we change the way we use it? Will we even allow discussion of how we harm and exclude people by our misuse of ICT and how to include them? (James 2:18-19).

We should object when people are forbidden to worship God. We need to learn to value our own religious freedom and what it means to be citizens of a free country. There is a difference between freedom to create and enjoy culture and living responsibly with the culture we create.

* * * * * * * *

". . . let us love, not in word or speech, but in truth and action" (1 John 3:18).

6. The Institutional Bias

The Bible and many governments support inclusion and oppose excluding people with impairments. Why then are many churches still inaccessible? Some churches meet in older buildings that are difficult to modify and some have limited resources, but in most churches that use ICT, it can be used accessibly. Something we may struggle with is the lure of the flesh and the allure of technology (1 Jn. 2:16). John Calvin (1846) rejected extreme asceticism, but he also warned against our becoming so delighted with marble, gold, and pictures that we become marble-hearted.

Self-centeredness, ignorance, and denial all play a role in ableism, that is, discrimination against people who have a disability. We think everyone is just like we are. We may not know people with hidden impairments, or not know how they function or are blocked from functioning by things we do and we hate to talk about, or even think about disability. We fear disability and deny it is a normal human variation. Therefore, we reject it as part of who we are and what we do.

We may imagine that it is our right to use ICT anyway we want, simply because we are familiar with it and like it. Or, we may fear that if instead of giving "the people" what they want, we were to present the compelling weight and beauty of righteousness, they would leave or not come to church. Or, we may be trying to attract a new, younger generation, but not realize that people with impairments exist in that younger generation too. We may not know restrictions on technology are common. How then can the church be moved to change how it uses its information and communication technology? We need to understand the many ways the Institutional Bias operates.

47

Turning Barriers Into Bridges

In a Biblical and Spiritual sense the Church is a living organism as the body of Christ (1 Cor. 12:27) and the bride of Christ (Rev. 21:9), not yet perfect, but making herself perfect (Rev. 19:7). The Church is God's family (Eph. 3:15), His temple (1 Cor. 3:16). Believers are becoming a Spiritual house and a holy priesthood (1 Peter 2:5).

In a natural sense, in terms of its physical plant, the roof that must be repaired and the water and electric bills that need to be paid, the church is an organization, an institution. As such, the Institutional Bias obscures individual responsibility and accountability under a guise of, "we had to do it," "we did it," or "we like it," or "they like it," or "they told me to do it," rather than, "I like it" and "I did it." It is often unclear who is really in charge. This spoken or unspoken claim of institutional authority, more than individual attitudes or ableism, are part of the less obvious, but more significant reasons churches are slow to change.

Many churches celebrate disability awareness month or week and call for inclusion. Christian media publish a sentimental story yearly to try to evoke inclusion of people with impairments and their caregivers, but the only messages presented are, "change the people." Rarely if ever is the message, "change the institution." Academic disability studies are acceptable, as are stories of, and pleas for more good deeds, or even a new theology—as if the old one were insufficient. A focus on changing people is permitted. Success stories of efforts to change individuals' negative attitudes about disability or to overcome social stigma are permitted. The emphasis is from Prov. 4:23 and Matt. 15:18-19, that what is in and comes out of the heart leads to evil behaviors. No doubt individuals need to be transformed, but our need for individual heart transformation should not overshadow attention on institutional level behaviors.

The Institutional Bias

The focus on individual acts avoids examining pervasive institutional acts or lack of action. Critics of culture can describe the ways the tools of mass media and the institutions of mass communication shape, control, and dominate culture–even Christian culture. But how disability discrimination works, or the message, "demolish institutional barriers," is all but censored.

The censored message is, "churches are harming people by their misuse of technology." Love does not harm its neighbor (Rom. 13:10). The only article explaining how to end disability discrimination in churches was published in one denomination's magazine 20 years after it voted to end disability discrimination in their churches. That long overdue article reported that most of the 800 churches surveyed were accessible to people who had a mobility impairment (probably due to local building laws), but only 50% provided some help for people with a hearing impairment and only 40% provided some access help for people with a vision impairment (Frank and Stephenson, 2013).

The Institutional Bias and technology has many facets and a very long history. It is as old as the telescope developed by Galileo Galilei in 1609, which helped verify that we are not the center of the universe--a revelation the church did not want then and some may still not want. Likewise, the printing press invented by Johannes Gutenberg in the 15th century allowed more people to access the Bible–but some in the church did not want more people reading the Bible then and perhaps today as well. Disability discrimination in church is maintained by self-centeredness, a lack of self-control, and a Christian outreach limited by our ignorance of Biblical, technological, and civic standards, as well as by ignorance, fear, and denial of the range of human variation. This operates by institutional behaviors, not only by individual attitudes and/or behaviors.

Turning Barriers Into Bridges

The loud silence surrounding technological barriers in church is as old as the Reformation which erupted due to the stance that the church is right and need not change no matter how much it strays from God's will or harms people. Institutions resist change, protect their unjust barriers, and reject their own culpability. Missing are calls to change institutions that by their misuse of technology disable people who have impairments. The acceptable messages refer to good things individuals do, or could do better. Even after 30 years of laws prohibiting disability discrimination, change is not urgent for institutions.

We prefer a person-to-person change process because institutions intimidate individuals. Then, insult is added to injury when an institution even attacks someone who protests unjust discrimination. This kind of barrier, denying the message by blaming the messenger, is not new. Ahab, the idolatrous king, called God's prophet Elijah, "the troubler of Israel" (1 Kings. 18:17). Jesus was called "Beelzebub" (Matt. 10:25). Typical labels today are, "ungrateful," "uppity," and "bossy."

People with hidden impairments may be ignored or may be labeled lazy, stupid, inattentive, or uncaring. But those who object to barriers in church may get attacked with labels used against protesters, such as, "heretic," "rebel," "complainer." and "trouble maker." Advocates who mention even obvious disability discrimination may be labeled, "divisive," "confrontational," or "impolite," "too pushy," "not diplomatic enough," or "not one of us," as an attempt to deflect their message. Messengers are warned of backlash if things change too fast, or they are encouraged to exhibit stoic endurance in the face of injustice, or warned against bitterness and offered as its cure, "love:" and "forgiveness," but without distinguishing between individuals and institutions, or between sinners and sin. David alone cannot battle name-calling bullies within Institutional Goliaths.

The Institutional Bias

Ryan (1971) calls such name calling "blaming the victim," when the wealthy blame the poor for their poverty. Kagle and Cowger (1984) and Rubin and Roessler (2001) use the term "blaming the client," when counselors or social workers blame clients for not improving despite their facing severe and intractable environments. Barnes (1996) notes how sources of funding influence scientists and college professors who thus can only blame, or mention problems with "the people," and not with the institutions that pay for their research and teaching. Sanchez and Fried (1997) assert that the right to expose the larger institutional context is a contest of power. In order to protect the status quo of the institution, those aligned with the dominant culture inhibit or block the seeking and sharing of information concerning censored aspects of the context.

The result, in the case of disability discrimination, is that mention of its meaning, or that it is illegal, or precise remedies, is absent or minimal in public media and in academic journals, textbooks, or training in Pastoral Ministry, Worship, Missiology, Theology, or in Ethics or Bible courses. Disability Studies and Disability Theology label disability discrimination a "civil rights" or "minority group" model, appropriate for study, which relegates essential advocacy to the ethereal status of an "optional activity." Disability remains defined as an individual's deficit rather than also being the result of institutions misusing technology. Then, the law leaves redress up to an individual with an impairment.

Churches promote charity, but may fail to first encourage justice in God's house (Micah 6:8; 1 Peter 4:17). They cannot sing with Philip Bliss, "Dare to be a Daniel, dare to stand alone, dare to have a purpose firm and dare to make it known." Discussion of ICT barriers in church is absent from pulpits, schools, or media and absent from our thoughts and prayers.

51

Turning Barriers Into Bridges

Discrimination on the basis of disability in churches is an institutional problem. A survey of 300 plus churches found that for 92% of the respondents, the pastor led the church into using computer projectors (Koster, 2005). This is a small sample out of the 350,000 public churches and the thousands of house churches in America, but it reinforces Roger's (2003) findings on the importance of opinion leaders for spreading innovation.

One respondent in Koster's survey mentioned being prepared to give large print paper handouts to the members who found they could not see the computer projection well, but during one hymn-sing, the song sheets provided were normal sized print in all capital letters, not large print text (which is 16 point font up to 36 point font and not all caps). If a church's leadership has not had disability accessibility training or has limited personal experience with impairments, they will be unable to guide their congregation into using ICT accessibly. A pastor, a worship leader or coordinator, a church secretary, and an Information and Communication Technology worker can usually read small text in print, on a computer, on the Internet, or projected. They may decide the church does not need or cannot afford 12 point Arial font as its "normal" size and "see" no need for some large print paper handouts and "see" no need for clearer projected text or clearer text on an LCD, because nobody mentions it and it does not impact them. They "see."

Over the past two or three decades we have used computers and sound systems incorrectly and we need to learn to use them correctly, inclusively. This needs to be consistently taught in churches, Christian schools, colleges, and seminaries and in two year colleges where ICT is usually taught. Its Biblical roots must be routinely mentioned as part of our worship and in church growth material and within a wide range of materials produced by Christian media, publishers, and organizations. Disability is not a separate topic. Disability and powerlessness-- the scandal of the cross–is a central message of the Church.

> To paraphrase Martin Niemöller: First they excluded those with low vision from church. I did not speak out because I could see well. Then they excluded those who could not read well. Again I did not speak out since I could read well. Then they excluded those who could not hear well and again I did not speak out because I can hear well. Then they excluded me, and no one was left to speak out for me.
>
> "If you were blind you would have no guilt, but now that you say, "We see," your guilt remains" (Jn.9:41).

Unlike the Holy Spirit, electronic light and sound effects are controllable and we control people with them. Institutions favor order, but the misuse of technology sends a message contrary to the Gospel and the Bible. It is also not an accurate reflection of society, our culture, or our people and it violates the law. Neither individuals nor institutions like being told they are wrong, yet Jesus rebukes those he loves and tells them to repent (Heb. 12:6; Rev. 3:19; Ps. 32:9). And Jesus, the Word of God specifically tells us to break the silence and rebuke the offender (Lev. 19:17; Luke 17:3). We miss the mark when we ignore the offence of the cross or boast in human power (1 Cor. 1:23-28; Gal. 5:11; Jer. 9:23-24). The next chapter briefly describes a "fear-of-making-waves" barrier we must overcome.

The "Institutional Bias" enforces the status quo, "just go with the flow." The "Majority Bias" believes "we're normal and you're not." It believes inclusion is extra and disability is rare even though one fifth of the population has an impairment and we are all only a second away from becoming impaired. People with impairments come in all ages and in every race and culture. Ignorance about impairments is not normal and ignorance about accommodations is often illegal.

The bad manners that result from the "Majority Bias" gave rise to several **Disability Etiquette** lists of good manners. Many of these lists can be found Online.
*   Don't ask about a person's impairment, whether mild or severe, without his or her permission to ask.
*   Don't endlessly discuss an accommodation as if it were some new and strange invention.
*   Refrain from unnecessarily sharing your likes or dislikes about another person's accommodations.
*   Include the topics of impairments and disability in conversations and Bible study, they are not unusual.
*   Include disability as part of worship, missions, and church life for everyone, not just as a separate issue.
*   Set up activities in an inclusive way, such as evangelizing with small and large print material.
*   Don't be condescending when providing help that is just our basic responsibility. It is not charity or pity.
*   Don't expect gratitude or celebrate the wonderful (but normal) accommodations provided in church.

Accessibility says, "Welcome Visitors," but hospitality is for guests or for strangers. Many people with impairments belong to our church or are our friends or family members. They do not need special or extra hospitality. They just expect appropriate respect.

7. Advocacy Beyond Style or Type

Most groups, Christian or not, will take care of their own when there is a need (Lk. 6:32; Gal. 6:9-11). Being accessible to the general public is always needed and usually achievable. Reasonable access is not a preference. Accessible, inclusive use of technology in church is providing something essential. Without it, participation in the Word, worship, fellowship, and Christian service is impossible. To paraphrase the Apostle Paul, "if technology makes a message unintelligible to some people, how will they know what is being said (1 Cor. 14:9)?"

Church Style: Each church has its own range of Bible exposition. There are Biblical topics besides the presence and perspective of people who have impairments that may be rarely dealt with or ignored. People with impairments, like everyone else, have a variety of interests and needs and desire various styles and teaching. They want to try out churches and choose where they will attend. That choice has been limited in some areas to where there are few or no Bible believing churches accessible to people who do not see, hear, or read well.

A church's ICT may be viewed as part of its purpose. The style may be contemporary, or traditional, or only upbeat, or allow some somber reflection. It may aim to be an outreach to unbelievers, or attract a younger, an older, or a mixed group. People who do not see, hear, or read well come in all ages. They need to be saved and nourished too. A small, large, or mega-church can be accessible. There is no style of church, not even a military chapel, a house church, or a small group, that must exclude people with impairments with its technology. If a church does not want in include people with impairments it may take the combined work of many advocates to open the doors.

Turning Barriers Into Bridges

Concert halls present rap, rock, or country music one week, and classical the next, but all accessibly. Even lip-sync can be accessible. Sports arenas, theaters, or zoos, medical facilities, or banks, schools, restaurants, or stores, and trains, planes, boats and buses must all be accessible and their staff trained to provide reasonable access. Why not our churches? Our message is the most important of all.

The purpose of our church technology and music ministry, as with any other ministry is, "to equip the saints for the work of ministry, for building up the body of Christ until all of us come to the unity of the faith and of the knowledge of the Son of God, to maturity, to the measure of the full stature of Christ." (Eph. 4:12-13). This typically means teaching Biblical, doctrinal truths through songs, sermons, and teaching sessions, or individual and group activities. This only works if what we communicate is accessible. It must be heard, seen, or read to be understood, learned and even memorized, hopefully practiced and by grace, lived out.

Most pastors, educators, and counselors understand that memorization, learning, and behavioral transformation occur and are enhanced by emersion and by experiential encounters. Many churches find that within an atmosphere of Holy Spirit inspired worship and Gospel proclamation people are drawn closer to God and eternal substance is imparted to them. Worship is an effectual part of equipping the Saints, any or many of whom may have or acquire a hidden, slight to severe, long-term impairment. If the goal is to attract seekers who are not yet Saints, that group includes people who have, or who will acquire impairments, who we could attract. If the goal is to grow like other churches, it is best to emulate a healthy, inclusive church. Whatever the group or the goal, we can be accessible.

Advocacy Beyond Style or Type

Whether in a more traditional church format with its emphasis on the words of worship, or in a contemporary style gathering, with its emphasis on experiencing the worship atmosphere, there are people who do not see, hear, or read well who could participate and contribute. They need not all be restricted, segregated recipients of a special outreach ministry. With better use of information and communication technology, more can partake of worship with the entire church. God is not the author of confusion, but of peace (1 Cor. 14:33).

Impairment Type: The Apostle Paul personally knew severe, long-term impairment (2 Cor. 12:9). He called for unity in the body because not all parts of the body function the same way, but all are needed (I Cor. 12:14-27). A portion of those who cannot see, hear, or read well may not read or sing regardless of the clarity or size of print, or may not be able to see, hear, or read at all. Some may simply not want to sing in church.

People with different types of impairments face barriers that must be demolished. At some time, everyone may need comfort, encouragement, and to be included, but our goals may differ. Moving to the beat of loud rhythmic music may allow participation for some, but harm or drive away others. Colorful computer projection with small print may attract some, but be a barrier for those with low vision or limited reading ability.

Despite these differences, more people, perhaps 95% instead of 68% will benefit from general changes that improve accessibility in church. There are measures that are easy to learn and to apply whereby we can include more of those whom we may now be excluding. We can extend God's grace, though we may not reach everyone, or even know who we reach and even though every area does not have the same level of need.

Turning Barriers Into Bridges

We read in Hebrews 12:13, "make straight paths for your feet, so that what is lame may >not be put out of joint,< but rather be healed." (>not be disabled,< in some versions), This fact, that environmental conditions can increase or diminish the severity of an impairment is a fundamental moral value of the Rehabilitation profession (Wright, 1981). It applies to physical and social conditions and became known as the Social Model of Disability. Laws prohibiting disability discrimination require that whenever reasonably possible we, (1) remove physical barriers, and (2) apply universal design, which means we make things and activities so that people who have impairments can use them, and we (3) provide reasonable accommodations (called adjustments in Britain). These steps diminish disability.

Unity of Advocates: Since those three steps cannot easily or equally help everyone, with any impairment, in all situations, the Social Model and the laws create two groups. One group is helped by the laws and changes to the environment, while the other group may need one-on-one help from volunteers or paid staff even in segregated areas. If that group feels the law leaves them out and fears making waves, it may instead prefer only nonoffensive, social interaction, and trying to change attitudes over advocating for accessibility changes. The terms "inclusion," or "person with a disability" for that group refers mainly to people with profound physical, intellectual, and developmental disabilities (IDD). Their caregivers may prefer seeking social acceptance, which they need, over the risk of alienating people if they advocate for inclusive projection, print, or sound, which they may not need.

The debilitating nature of the fear-of-making-waves barrier is highlighted in John 9:19-22. Synagogue leaders asked the parents of the man born blind to verify that the one Jesus helped was their son and to tell them how he was healed. His parents were afraid they would lose their position in the community so they owned their son, but did not own Jesus.

Advocacy Beyond Style or Type

Their son confessed Jesus was working in his life and for confessing that fact, he was cast out. Fear of loss of community or loss of needed assistance due to one's own or a family member's impairment still divides disability advocates.

For both groups, advocates are involved. Often it is the parents, not the person who has a disability that have a strong, well-known voice, but they may only know about the needs of those they care for. People with various types of impairments benefit from the combined efforts of many advocates even when they have different goals, especially if the loudest, or the only voices heard are the caregivers or parents of people with only certain types of impairments. A fear barrier can keep advocates silent, separate, and on opposing sides of issues.

The caregivers of people who have intellectual, or developmental impairments, or who have multiple profound disabilities who are not helped by accessible technology may not know about technology barriers. They, or their son or daughter may not want to read along, or sing along anyway. Nevertheless, advocates for people with disabilities not affected by the way a church uses its technology could support the efforts of those who do care about accessible technology.

We are united by a Gospel that changes insensitive hearts and by the fact that incorrectly used technology may create barriers to people who have a vision, reading, hearing, intellectual, emotional, developmental, or physical impairment. Some people with any type of disability in a church of any style are helped by accessible use of information and communication technology. We do not have to choose one group over another. We can be advocates beyond style or type as followers of the Messiah who helped people see. hear, and understand the Word of God, who, on the cross, was disabled and died for all.

Turning Barriers Into Bridges

Accepting people with impairments, just as they are, without demanding they be fixed, requires that whenever we can, we make our physical and social environments accessible. This includes how we use technology. It will not help everyone. Some would not want to come to church no matter what we did and a small percentage could not see, hear, or read no matter how we use our technology and some spaces and technology cannot reasonably, easily be changed. This occurred in the first century too. At that time in the Holy Land, well before glasses were invented, many people who did not see, hear, or read well were called blind, deaf, illiterate, or uneducated.

Removing barriers and using universal design and providing reasonable accommodations and adjustments, all without undue burdens or altering our essential purposes, can help some people with any type of impairment and in any style of church. It is not a question of us or them, here or there, or helping or harming this or that group, or trying to help only our own, or only the neediest. As this oft used play on Frost's poem says, "Two roads diverged in a wood and I took them both."

Consider how King David dealt with the exhausted third of his army after God gave him victory following the destruction of Ziklag (1 Sam. 30:9-25). The army rescued all the captives and recovered far more spoil than anyone had lost, but among the two thirds who were strong were those whom the Bible calls corrupt and worthless fellows. They did not want to share the blessings God allowed them to acquire with the exhausted one third who were not able to keep up. David made it a law in Israel that all in his army, weak and strong, would share alike. We too can reach for that goal in our churches.

* * * * * * * *

The next chapter lists more barriers we need to bridge.

8. Am I My Brother's Keeper?

Cain refused to admit he had caused harm or that he was responsible for the harm he caused (Gen. 4:9; 1st John 3:12). His testimony illustrates two facets of denial that keep pulpits and classrooms silent on this issue despite the rich Biblical record concerning people with impairments and despite the laws prohibiting disability discrimination which are the context and environment of our churches. In addition to ignoring communication barriers we create with our technology or the harm we cause with it, the following are some other, often repeated reasons for not being an accessible church.

(1) The technology myth says a technological fix or cure will come if we wait. Something will turn up so the majority will not need to change its ways for the minority. Maybe we can e-mail each person the song lyrics, sermon outline, or bulletin so each one can print his or her own pages in whatever size text desired. Some new, do-it-yourself technique will emerge!

Unfortunately, computers and the Internet today are becoming much less accessible for people with impairments (Frank, 1999; Frank, 2006; Blanck, 2014). More barriers to Web equality are being faced by people with disabilities even when solutions exist. "Let them take care of themselves" sounds as easy as bringing your own earplugs to church, but it is not easy and it does not encourage our meeting together (Heb. 10:25). The message sent by a "make-them-do-what-no-one-else-has-to-do" approach is, "stay home!"

(2)   "Send it to the committee," or, "do more research," are well-used ways to kill or delay action on something that can begin immediately. This ruse claims that there are no clear

problems and solutions. The scriptures and the research on disability discrimination are plentiful and conclusive, but rarely if ever shared in church. A committee may need to start by learning the difference between a special ministry and applying self-control and obeying the law when using technology.

(3)   One way research is used to block access in church is to survey a congregation to see if enough people want it. Besides ignoring the larger community as well as the future, church members might not say they need help in order to avoid conflict or they may not know how easy it is to be accessible. Another misuse of research is to aim only for the young because a survey found about half of Christians said they were converted at a young age. This overlooks the record in the Bible and in church history and excludes the many saved as grown adults, and those adults still needing to be reached and fed. Research is also misused when church members with a hearing loss are asked to vote on a hearing loop, but they have no experience with any type of hearing assistance system and are given an exaggerated price quote. Not surprisingly, it will be voted down. We may falsely think an uninformed, misinterpreted, or slanted survey is an attempt at research, or democracy, or we may believe the majority rules, but in a modern democracy where the majority rules, the minority is protected.

(4)   Another barrier is to agree to the easy and obvious, but delay using technology inclusively until volunteers or staff have training and a comprehensive plan to deal with all accessibility needs, even the most difficult or expensive. This is avoiding the easy by aiming for the hardest. A wiser approach is found in Proverbs 3:28, that is, do what you can do when you can do it. Do not wait. The methods for using technology accessibly have been available for decades and are easy to apply now. The Biblical perspective in chapter 2 and 3 can be shared now.

(5)   A related barrier is to consult retailers concerning ways to improve the sound or computer projection. They examine the facility and suggest buying a better sound board for $10,000, a stronger computer projector for $6,000, or an LCD screen. Then, the false conclusion is given that accessibility is too expensive. Retailers may say the problems can be reduced with curtains or sound absorption material, but they are unlikely to suggest lowering the pitch of music, or increasing font sizes and spacing, or projecting fewer words on each slide, and improving the letter/background color contrast of an LCD or projected text. They may not be aware of those steps. The lure of selling and buying new technology makes a new computer and a plug-it-in-and-push-one-button solution attractive. Unfortunately, this may only lead to a brighter and louder church, but not greater clarity.

(6)   A sad obstacle to an accessible church is forgetting that God shows no partiality (Acts 10:34-35; Ro. 2:11). The idea that nobody here is disabled, that this is for other people, those other people somewhere else, is unbecoming a Christian church. The prevalence and type of impairments vary between regions, races, income levels, gender, and age groups, but limited vision, hearing, or reading ability affect various types of people everywhere. The Bible says, "There is neither Jew nor Greek, there is neither slave nor free, there is no male and female, for you are all one in Christ Jesus" (Gal. 3:28).

(7)   Sadder still is the fear that valuable relationships could not or would not develop even with better access in church. In response to that objection, I have to confess that I too have participated in conversations at church with no personal or spiritual content. It is much easier to talk about topics commonly heard in secular environments than to engage others in deeper spiritual or personal issues. I can recall church suppers where

each table was filled with only one family and there was no mixing at all. People with impairments were not the only ones left out or segregated. I can think of small groups that discuss their latest work or family news, but no longer find Bible study appealing. I also know first hand the chill that can and does come into relationships when disability discrimination and equality of access in church are mentioned.

Few of us are popular, extroverted leaders, whether able-bodied or not. We mostly just try to fit into the social structure around us, rather than create or change it. But Jesus said, "By this everyone will know that you are my disciples, if you have love for one another" (Jn. 13:35), and "it is more blessed to give than to receive" (Acts 20:35). Note too, "A man who has friends must be friendly" (Prov.18:24). Access is a precondition for participation in church. It is not the end point. All of the Gospel must take root and grow. Jesus said, "if I be lifted up, I will draw all peoples to myself." (Jn. 12:32, NKJV). As we draw closer to Jesus we will draw closer to each other.

As I share on this issue, these and other deflections to being an accessible church are repeated in various forms. Ignorance of the problems or easy solutions may be an issue, but barriers may exist because of underlying philosophies and personal motives, including the fear of an impairment and its potential for dis-ability. One church decided to do nothing until after they remodeled their sanctuary which was unrelated to the changes needed. One group declined because they only knew of about two percent in their church it affected, 8 out of 400. Eight people were in Noah's ark. Even opponents of abortion and euthanasia may fail to see the contradiction between their pro-life beliefs and their actions when they deny people with impairments access to church by misusing their ICT.

Denial and avoidance disguised as religion can be difficult obstacles. As a hospital chaplain I would sometimes encounter people who insisted on success only as they defined it. They denied disability is, or can be a normal part of life and refused to believe any severe impairment could happen to them or to their loved ones. There were also those, both patients and their family members, who deflected every question concerning disability or their path forward. They could not or would not deal with the details of living with an impairment, even though the patient could not be released without a safe place to go.

For many, a long-term impairment is the thing we fear the most (Job 3:20-25). We project our fear onto others and assume that they, like Job, are miserable. The truth is, many of us with a long-term impairment live fairly full lives--except for the disability discrimination we face in churches and elsewhere.

Another barrier is the anti-law (antinomian) position which exalts grace and freedom as the right to do whatever one pleases or thinks profitable, rather than as the opportunity to do what is right. Lawlessness ignores the Bible's view and invents fault with laws that restrict harmful behaviors. Resisting government intervention in churches may at times be justified, but not as an excuse to ignore secular regulations of ICT while at the same time we imitate secular excesses with ICT.

Access gets denied by "dancing around" the issue, or by "polite silence," or by denying accessibility standards exist, or denying medical, educational, or legal standards exist, such as, the range of variations, slight to severe, between some vision and total blindness or hearing and total deafness, or the ranges from illiteracy to literacy. Confusion is created by asserting that generic standards might exclude individual needs, or that they are only subjective opinions, or that every limitation or only the most extreme conditions are an impairment or a disability.

Another confusion is framing disability as a contest between winners and losers in a competition, as if in a sporting match. This rejects the notion of fair play and reveals a "might makes right" ethic that leaves us back in Judges 21:25, where everyone did what was right in their own eyes. Even opponents of evolution may embrace social Darwinism and survival of the fittest, saying, those who are unfit should not be in church.

The belief that computers are easy to use and yet fearing the process of learning to use them accessibly is yet another contradiction we live with. Also, our perseverance and frugality are tested when nobody uses the large print we provide in advance of requests, but regaining people's trust takes time. The goal is for us to cultivate the spiritual fruit of self-control in an environment where access for people with impairments and knowledge about providing access is required by Biblically based, but poorly enforced, secular laws. Who but Satan would want to hinder worship to God (Isaiah 14:12-15; 2 Cor. 11:14)?

Another deflection is the idea that disability is all about voluntary, individual relationships, or is just another "cause" for charitable activity. It restricts our focus to the people we help, or who need our help, or to the good deeds we do and evils we avoid, or to other people we also unnecessarily exclude. This frames **the disability we cause by misusing ICT** as one more choice or interest group among many, but disability discrimination is a sin and illegal, and is easily avoidable. And the list goes on.

Determining the exact number of people who need inclusive ICT is another obstacle. It asks, "Who is my neighbor?" instead of asking, "To whom am I being a neighbor (or not)?" Nonetheless, to learn more of what it means to be human and as promised in the introduction, the next chapter offers disability demographics for America, Britain, Canada, and the world.

9. Who Is My Neighbor?

Demography is an old, imperfect science because people are involved. We prefer easy-to-read data. Social science researchers know all too well the expenses and difficulties of collecting and reporting data on people with impairments. We use what are considered the best data without claiming all-knowing omniscience and there is room for improvement.

Presenting multiple data sources may seem confusing, but there is not one easy number to report for each population. Several are given here in order to acknowledge the differences in the reported dimensions of these populations. Differences may be due to the intended purpose of the data, or due to the questions asked, or the definitions of disability used, or the range of ages considered, or other characteristics of survey respondents. The results from different data sources do not question the existence of large numbers of people of all ages with seeing or hearing impairments, or limited reading ability.

America: The American Foundation for the Blind (AFB, 2012) estimates that 20.6 million Americans cannot see well even with glasses or contacts. The National Institute of Literacy (NIL, 2013) estimates that 40 million Americans of working age find even a basic literacy level (11 year old, or 5th grade level) difficult. The Hearing Loss Association of America (HLAA, 2014), citing the Center for Disease Control data, estimates that 48 million Americans have some degree of hearing loss. These estimates cover impairments ranging from slight to severe including the smaller percentage who cannot see, hear, or read at all. Inclusive use of technology in church could help many, but not all of these people. Not all of them would want to attend church and some are members and already attend church now even though they cannot fully participate.

Turning Barriers Into Bridges

Some surveys do not specify the population parameters. Some surveys omit people living in institutions and some refer only to the 177,295,000 people between the ages of 21 to 64, not the entire U.S. population of more than 308 million (see census.gov 2010). This omits 130 million people. Based on the Survey of Income and Program Participation (SIPP), out of a population of 308 million, 16.6 percent (or 51.6 million) had disabilities, of whom 6.6% had difficulty walking, 6% had mental disabilities, 5% (or 15.4 million) had difficulty seeing, and 1.9% (or 6.2 million) had difficulty hearing (Brault, 2012).

Consider some other sources. The National Institute on Deafness and other Communication Disorders (2005) estimates that 30 million people in the U.S. aged 12 or older have hearing loss in both ears (not 6.2 million per SIPP or 48 million per the CDC). The AFB cites the National Health Interview Survey data that 20.6 million people have difficulty seeing even with glasses or contacts (not the 15.4 million per SIPP). The estimates differ, but it is not always clear why. The definition of an impairment, or year of a survey, or the ages included would effect results.

Large variations, a low or high estimate, do not mean the information is invalid. We choose data sources for a reason and we include people in church for a reason. If employment is the purpose, ages 21 to 64 make sense. If church participation and respect for parents and for the elderly are the purposes (Eph. 6:2-3, Lev. 19:32; Ps. 71:9; Is. 46:3-4), then we ought to use data that include all ages, not only a part of the population. The combined incidence of all disability types is three times higher for people over the age of 65. Are people over 65 being left out of churches? Are we honoring them? They are living longer and in some areas they outnumber those under the age of 18. Furthermore, disability exists among younger people as well.

Who is My Neighbor?

Gallup polls (2010) typically find that roughly 40% of the American population say they attend church. The numbers vary by region. The exact number of churches in America that misuse ICT is unknown. The precise number of people of any age who cannot participate in churches that misuse technology is unknown. Survey and anecdotal data suggest that. 60% to 75% of the 350,000 churches in America misuse ICT and unnecessarily disable and exclude up to 80% of people who have impairments and their families. When differences or uncertainty in data exist shall we abandon compassion and act contrary to Abraham and say, "Lord, we will not be inclusive because, peradventure there be not 50, but only 45, or 40, or 30, or 20 or 10," (Gen. 18:24-32). What if there was only one more person we could include, instead of millions more?

Britain: Just over one third (34%) of adults living in Britain said they had a long-standing illness or disability (LSI) and almost one fifth, (19%) said they had an LSI that limits activity (Office of National Statistics [ONS], 2012). With a total population of slightly over 64 million, there are almost two million people in the United Kingdom with sight loss (National Health Statistics [NHS], 2013). The number is expected to increase to 2.25 million by 2020 and double to four million by 2050 (Stade, Royal National Institute of Blind People, [RNIB], 2014). About 3.5 million people of working age (16 to 65 years) are "deaf or hard of hearing." However, almost ten million people in the UK have "some degree of hearing impairment or deafness" according to Action On Hearing Loss (2014). Around 16 per cent of adults, or 5.2 million are described as "functionally illiterate," that is they have literacy levels at or below those expected of an 11-year-old in the 5th grade (National Literacy Trust, 2011).

Turning Barriers Into Bridges

The combined data for adults between the ages of 16 and 65 in Britain indicate slightly more than 17.2 million reports of a vision or hearing loss, or low reading ability. These could overlap with some individuals having more than one sensory impairment and/or limited reading ability. The number could be nearly twice that if all ages were included. The number of immigrants in the UK is large and illiteracy is more often an educational deficiency rather than a learning disability. Some immigrants are from countries that deny education to females. Again, this is not 23% of the entire population of Britain. It is neither 17 nor 34 million individuals, but rather, 17.2 million, or, if all ages were included, twice that many reasons to use information and communication technology in accessible ways in all of the 50,000 churches in Britain.

Canada: In 2012, with a population of more than 34,752,000, 3.8 million adult Canadians age 15 and older, or 13.7% of the adult population, reported being limited in their daily activities as a result of a disability. This includes 3.2% of the adult population having a hearing loss and 2.7% having a vision loss (Statistics Canada, 2014). In Ontario, with a population of 13,135,000, 1.8 million adults report having a disability.

The Canadian Association of the Deaf (CAD, 2012), with strong disclaimers as to the dependability or accuracy of any data, reports 350,000 profoundly deaf and deafened Canadians and possibly 3,150,000 who are hard of hearing. The Canadian Hearing Society (CHS, 2013), reports nine million adults (over one-fourth of the population) have some hearing loss, with one million having a hearing related disability. According to CHS, some studies indicate the true number may be three million or more adults with a hearing related disability because those with hearing problems will often under-report their condition.

Who is My Neighbor?

There is a substantial difference between "some hearing loss" in nine million adults and an "under-reported hearing related disability" in three million adults and reports of "a hearing related disability" in one million adults and 350,000 who are "profoundly deaf." The ages included, the range or definitions of what is measured, and the accuracy of the responses all affect survey results, but the numbers are still huge. The 2012 Canadian Survey on Disability (CSD) conducted by Statistics Canada found more than a million people (1,137,292) with a hearing loss in Canada with 437,718 in Ontario (Employment and Social Development Canada, 2015).

The CNIB (2007, 2013) reports that half a million Canadian adults have a vision loss. It is not clear if that means (in the same way as the hearing loss data), "some loss," or if it means enough loss to be at a level of a "vision related disability," or if it means total or near-total blindness. This data may also be incomplete due to under-reporting.

On the other hand, according to the 2012 Canadian Survey on Disability (CSD) by Statistics Canada, there are almost one million (959,590) people with a vision loss in Canada, including over a third of a million (369,324) in Ontario (Employment and Social Development Canada, 2015) or almost twice the number reported by CNIB in 2007. In addition, for the first time in Canadian history there are more people who are age 55 to 64 than age 15 to 24 (Statistics Canada, 2014) and as already noted, while the incidence of disability increases with increased age, impairments exist in people of all ages.

It is understandable that people would under-report a hearing and vision impairment. People who do not hear or see well might not respond to a spoken or written survey consistently. Problems exist in researching and reporting the demographic data on both the incidence and severity of impairments. The sources of the data, the Participation and Activity Limitation survey (PALS) and the Canadian Survey on Disability (CSD) are not equivalent surveys and cannot be compared, but the highest and lowest counts are large.

Comparing the data from different countries and within the same country is made even harder because the definition of adult varies in the data, (15 to 65, or 21 to 64). Also, the cost and methods of surveying and testing and the meanings of the terms "loss" or "disability" vary. Furthermore, the consequences of being labeled as having limited hearing or vision vary between countries and within countries. There are incentives for under-reporting, such as, the fear of losing privileges, or a lack of benefits or helpful medical or rehabilitation services or supports. Still, the data indicate a large population with hearing and vision loss who could not fully participate in any of the 30,000 churches in Canada, including the 11,000 churches that consider themselves Evangelical, whenever the churches misuse information and communication technology.

There are also huge numbers of people who do not read well in Canada. In 2003, 42%, or 16 million adults, age 16 to 65, had low literacy skills (Canadian Literacy Network [CLN], 2015). In 2012, almost half, 48.5% of Canadians, age 16 to 65, had low literacy skills with scores in the Level 2 category or lower; that is, a 5th grade or 11 year old level or lower. In the later primary grades tracking, scanning, and fluency can all be adversely affected by smaller fonts and by increased concentration of letters, words, and lines on a page. In Ontario, 47%, or 6.2 million adults had low literacy skills (Employment and Social Development Canada, 2015). Canada is a bilingual nation with a high incidence of immigration. The reported decrease in literacy may be largely attributed to increased immigration from countries where the opportunity for an education is limited. Those factors are taken into consideration by CLN which offers reading tests in more than one language.

In 2011, according to Indicators of Well-being in Canada (2012), disability services were needed by 12% of the college students in Ontario. As of 2012, almost 25% of adults in Ontario have a university degree and almost half of the adults in Ontario cannot read well. Whatever the reasons for this education gap, there is a huge need for large print, and accessible projection of Christian materials, and truly large print Bibles.

There is little point in arguing over demographic data as though making a cost/benefit estimate. We can instead use the best standards for readability without trying to simplify or dumb down the material. Church communities consist of people of all ages who have a variety of strengths and weaknesses. We cannot assume any group, young or old, does not need to, or does not want to participate in church. Our information and communication technology today allows us to easily create larger print, better color contrast, better letter, word, and line spacing, and fewer words per page or slide. Can we lower the pitch and add a hearing loop or FM system to our sound systems because one person needs it? Jesus asked, "Which of you will not leave the 99 to go find the one?" (Matt 18:12).

The main issue is how we use our technology each week even when no one is around except able-bodied people. Misuse is systematically occurring as churches are set up every week. People with impairments of any age may already know from past experience not to come to church. They only have to look around and listen to know if a place is accessible. They are not likely to tell us about their hidden impairments. We have our work cut out just to get them to visit again. Our missionary outreach may also be unnecessarily inaccessible.

Worldwide: The World Health Organization (WHO) of the United Nations estimates that 20% of the world's population of seven billion has a severe impairment. If all 1.4 billion were in one country, it would be the third most populous nation after China and India. They do not all have a vision or hearing loss. Based on countries which collect such data, 314 million people worldwide have impaired vision and 360 million have a disabling hearing loss (WHO.org, 2010). In addition, 775 million adults worldwide lack minimum literacy skills (UNESCO.org, 2014).

Language translators work for years creating a Bible and Christian literature in the native tongue of an unreached people group. But then, the technology that could produce various print sizes is often used to reproduce only one size of print books, or papers, or on a video, for a people with limited or no access to optometrists or reading glasses. Audio material and internet sources help some, but they cannot replace books and paper.

From Roman roads to satellites, advances in technology have opened doors for sharing the Gospel and they can do so even more today if we use our missions resources inclusively. The key is to consider how our technology can be used as a bridge to include 95% of the variations in the population, not just 50% or 68%. We can plan to be barrier free in the first place. We need to learn and teach that not everyone is an able-bodied, highly literate student, teacher, minister, or missionary.

The focus needs to be on building our environment and using our tools in accessible ways—not on the number of people with impairments—and not only one at a time. A self-evaluation, accessibility audit can reveal what we are doing wrong and guide our corrections. Thirty years ago we had fewer choices. We were limited by print books and typewriters. Today we have the technology to set up church and outreach ministries in easy and inexpensive ways that could reach many more people.

Who is My Neighbor?

The moral fiber and character of a nation come from the Bible and from the churches. We can lead and be a light in the world rather than waiting until others force us to do what is right. Most of us object to being told we are wrong, but before asking, "What extra thing can we do to help those we inadvertently left out?" we can first ask ourselves, "Are we being a neighbor in the social, physical, and technological environment we create?" We are called to help others, but we can also avoid harming others in the first place by considering how we can use our technology more inclusively all the time.

There are limits to what we can do. Impairments range from slight to severe. The generic accessibility standards listed in the next chapter will not help everyone and individual help when it is available will still be needed, but it is not necessary or normal to be cut off from Christian community because of limited vision, hearing, or reading ability. When a church is inaccessible because, "we didn't know," or "nobody here needs that" (yet), it is likely that people with impairments who were ignored did not say anything. They just participated less or left. They may have chosen to leave for a variety of good reasons, but if disability discrimination was a reason people left, we may have chased them away and we have work to do to bring them back, to reconcile, and to restore. Hopefully, when people learn that a church is trying to be accessible, the community will help (1) remove barriers, (2) create universal design, and (3) provide reasonable accommodations and adjustments.

Bible believing churches can lead the way in being accessible because this is central to the Gospel, it is not something extra we should do. Those who rebuked Bartimaeus changed their tune when they heard Jesus call for him (Mark 10:46-52). The next chapter shows how we could change too, if we have ears to hear what the Spirit is saying to the Church.

Beware of Geekspeak: If someone claims the church computer's CPU, and RAM, or its hard drive memory are too small or too old to project or print large print slides or paper, point out the weekly super-large print slide, "Offering Time" or the huge computer created paper sign in the entranceway, "Support our Missionaries." Small or large print, or projected or LCD slides can be produced on almost any computer. When a new computer is bought and the choice is again made to use a church's ICT in only inaccessible ways, remember, "it's a poor worker who blames his tools for his mistakes." (See also Exodus 32:24.)

"We who are strong ought to put up with the failings of the weak, and not please ourselves. Each of us must please our neighbor for the good purpose of building up the neighbor. For Christ did not please himself" Ro. 15:1-3.

At around age 45, many people experience vision changes and need glasses with bi-focal or tri-focal technology. This is a normal aging process and is not usually considered a vision impairment. Reading smaller print is possible with those tools, but the eyes and body may need to be carefully positioned and held steady to read. This can be tiring, even painful over time. Larger text can help.

10. A More Excellent Way

For many churches, computers and sound systems are recent additions of only the past two or three decades. We can change how we use them. The best access is not based on opinion or preference. Choices exist, but the most readable formats for computer printing, photocopy, signs, projection, or LCD screens have been researched over many years. So too, the best accessibility for, the Internet, Web sites, e-mail, and other social media, smart phones, smart boards, TV, and video is known. A variety of hearing assistance devices are available as are accessibility features and devices for sound systems.

The following guidelines are not meant to suggest a church begin to use technology or upgrade what it owns. A church can apply these accessibility standards to the ICT it already owns, with little cost or effort, at a reasonable pace. The more expensive items mentioned are a hearing assistance system added to a sound system, or a Braille embosser added to a computer. Those may cost as much as a computer system, a computer projector, or a large LCD screen.

These standards come from various sources such as the American Foundation for the Blind (afb.org), the American Printing House for the Blind (aph.org), the Royal National Institute of Blind People (rnib.org) of England, the CNIB (cnib.org) of Canada, the World Wide Web Consortium (www.w3.org), the U.S. Access Board (www.access-board.gov), and also the U.S. Justice Department (www.ada.gov). Most manufacturers do not mention how to use their products or programs accessibly, but they might answer phone or e-mail questions on how to do something specific, like increasing font size, or letter and line spacing. Many libraries and colleges will answer questions about accessibility or how to use a program.

Turning Barriers Into Bridges

These lists are guidelines and standards for information and communication technology. They are not offered as legal advice, or as training in using software, printers, projectors, or other technology. An Internet search may help find how to use a particular program, or find an advocacy group, or accessibility standards for people with other types of impairments. A 53 page, introductory, general overview, discussing the inclusion of people with a variety of other kinds of impairments, (Davie, 1997, 2016), can now be downloaded from, www.aapd.com/wp-content/uploads/2016/03/That-All-May-Worship.pdf.

The following lists of precise standards will help many people who do not see, hear, or read well, but It is a good idea to ask individuals what they find useful, because people vary and impairments range from slight to severe. Other barriers may exist or emerge, such as lighting, seating, or architectural issues, or the introduction of newer technology. Signs, such as bright color strips on the edge of stairs, a stage, on an incline, or over electric wires connecting ICT, are usually mentioned in regard to accessible architecture, which needs its own book.

These tried and tested reasonable accommodations, required by Biblically based laws, enable people who have impairments to participate in our feasts of worship and the Word (Luke 14:13). Churches, as non-profit, authorized entities, can create, save, and distribute accessible formats. Again, this is not legal advice. See the 1996 Chafee Amendment [17 U.S.C. § 121], (www.loc.gov/nls/reference/guides/copyright.html), and see Article 2.c of the signed and ratified, 2016, Marrakesh Treaty to Facilitate Access to Published Works for Persons Who Are Blind, Visually Impaired, or Otherwise Print Disabled, at the World Intellectual Property Organization (wipo.int,) site, (www.wipo.int/meetings/en/doc_details.jsp?doc_id=241683). Learn about the Marrakesh Treaty at (www.worldblindunion.org).

A More Excellent Way

1) Printed Material: Computer programs, templates, and printers may automatically compress and reduce font size and type, or letter, line, and word spacing. It helps to have an accessibility quality control edit before and after printing.

Font size and style:

- Large print text is defined as 16 point font by the RNIB, 16 to 20 point font by CNIB, or 18 point font by the AFB. 18 point = 1/4 inch, 72 point font = 1 inch. Avoid fonts that are condensed or if 18 point does not = 1/4 inch.
- Average print size is 12 point font, but too often in thin Times New Roman font. Use a wider, sans serif font.
- The RNIB suggests a minimum "clear type" standard of 12 point Arial font as "normal," (not large) text.
- It is difficult to see and read bulletins or other material in a thin font style or in less than 12 point font size.
- Readability is best with only one style and size of a plain sans serif font, such as Arial, or a larger font type such as Verdana, without adding appearance affects or condensing or compressing letters, words, or lines.
- Expanding letter, word, and line spacing to 1.1 may be helpful. Aphont, an expanded font created by the American Printing House for the Blind (aph.org), can be downloaded for people who do not see well.
- Some sources use the term "larger font" for 14 to 16 point font and use the term "enhanced large print" for 24 to 36 point font (Kitchel, 2004).
- If 36 point font is not large enough, audio or Braille is recommended. Large print will not help everyone.
- Some sources suggest saving anywhere from 1% to 30% of printing costs by using smaller, thinner fonts, but this does not save money or souls.

79

Formatting:

- Reading is easier if each large print page contains meaningful material that makes sense rather than having partial sentences or bits left-over on a page.
- Instead of using a new page or larger sheet of paper when only a few words or lines are left, reduce the size of margins, or reduce the size of large headings, or eliminate unneeded blank space.
- An 8½ by 11 inch sheet of paper filled by 12 point text can be reformatted into 18 point, Arial font and fit onto an 8½ by 14 inch sheet of paper.
- Large print handouts can be created and folded so they do not look or feel bulky or obvious. If large print materials look or feel odd, they are not likely to be used.

Color and Contrast:

- The best contrast for printed material is black or dark ink on yellow, white, or very light pastel paper.
- Do not use green, blue, purple, red, or gray paper.
- It is harder to read dark ink on a dark background or light ink on a light background. Light on dark may work.
- It is harder to read text on paper that has pictures, designs, patterns, multiple colors, shading, or lines.

Other considerations:

- A few or a few dozen 18 point or larger bulletins, newsletters, outlines, or song sheets on 8½" by 11" (or 8 ½ by 14" or 8 ½ by 17") folded paper are not costly.
- Slightly larger print may help everyone, but true large print is needed by 10% or more of the population.
- When distributing tracts, the Gospel of John, or other literature, some material, at least some of the essentials and contact information should be in large print.

- If creating 100 bulletins each week is typical, begin with 90 in 12 point Arial font and 10 in 18 point Arial, or 100 in 12 point and 10 in 18 point. Add more as needed.

- It may take time before people become aware of and comfortable using large print materials. They need not be wasted. Anyone can use the large print copies.

- Provide paper copies of projected material to people who might not standup, but who may have their view blocked by those standing in front of them.

- Unless we ask or talk about it, people may assume their squinting and pain are just their own problem, even though it may be caused by our misuse of technology.

- With a TV, video, or DVD presentation, include a reasonable amount of readable, captioned text and audio narration. It can help everybody.

- A church library can have a large print section with Bibles, hymnals, devotionals, and other Christian large print books, such as those published by Gale Cencage, Walker, and other publishers that adhere to the standards of the National Association for Visually Handicapped (NAVH) which since 2010 is part of the Lighthouse Guild. Christian audio books are also a good addition to any church library.

Digital software systems with Bibles and reference works which have text that can be enlarged include the free Bible software program, e-Sword (www.eSword.net). Many types of digital books are available to download for free from the Gutenberg project (www.gutenberg.org/catalog). Bible software, such as Logos (www.logos.com) and WordSearch, (www.wordsearchbible.com) are for sale. The book texts can be enlarged, but the controls to use them may be small. Kindle (mobi) and Nook (epub) books have only limited accessibility.

2) Signs: Signs on roads are made to be readable. Signs in grocery stores are large and clear. Signs in churches point to a more important destination and a greater nourishment.

- Text in signs such as door numbers over a door or at a similar distance from the reader should be 3 inches tall by 1.8 inches wide (7.62 by 4.57 cm), with light-colored characters on a dark background.
- Event and information notices on a wall or bulletin board should be in large print with nothing blocking a person's approach to the notice for closer reading.
- A name tag is a sign. Use the largest, clearest text that fits, with the best color contrast, on the typical paper inserted into a 2.5 by 3 inch (6.35 by 7.62 cm) plastic name tag, or whatever size name tags everyone wears.
- For guidelines for Braille and other tactile signs see www.access-board.gov.
- Secular laws and common sense require that signs be easily readable for emergency exits, washrooms-lavatories, fire extinguishers, stairwells, elevators, or a defibrillator unit. Those are not new requirements.
- Some churches list their accessibility features on a Website, or in a visitor packet, and/or have a sign or some other large print handout to let the public know the various ways the church is accessible.
- Readable signs help locate: large print or Braille hymnals, Bibles or audio devices on a shelf, a ramp behind closed doors, or a wheel chair in a closet, or large print song lyrics, bulletins, or sermon outlines.
- Let people know what is available and the location, and make extra effort to announce if, or when things are moved, or items are changed, added, or missing.

A More Excellent Way

3) Bibles: We have a wealth of Bible versions. People who seek to know God need a Bible they can read.

- Bible publishers use their own definitions for the terms, "large," "extra large," "giant," or "super giant print," often without indicating the font size or sizes used. Their terms and font sizes may not be standard large print.
- Bookstore employees unaware of these print size variations in the industry may inadvertently mislead customers and some mislabeled items must then be taken or mailed back, or cannot be returned at all.
- The Lutheran Braille Workers produce large print and Braille Bibles. The Jewish Braille Institute distributes the Torah in Braille and in large print. Both, however, may be overwhelmed with requests.
- A complete, one volume, 18 point font, true large print Bible, King James version (KJV), is available from the American or Canadian Bible Society (ABS or CBS), or Holman, or Dake. Holman sells an 18 point font New King James version (NKJV). All of these would be easier to read if the letter, word, and line spacing were increased, but they might then be larger and heavier.
- Zondervan offers a one volume "super giant print" New International version (NIV) but it is only in 16.5 point font.
- Cambridge University Press publishes the KJV, the NIV, and the New Revised Standard version (NRSV), each in four volume sets, in 18 point font, which all have good letter, word, and line spacing. Thomas Nelson, sold a very readable, 24 point font, three volume KJV.
- ABS sells the easy to read, Today's English version, New Testament, Psalms and Proverbs, in large print.
- Additional complete versions, in true large print, in one or more volumes, would not be costly to produce.

83

- Bible software and eBook devices allow for changing font size for the text, but they also need their controls enlarged and better contrast, color, and lighting control. EBooks have limited accessibility even on computers, but Online HTML5 books are more accessible. Alternatives to reading, such as, audio Bibles, tape or CD exist, but to hold and to read a print Bible is often preferred over listening, or using a computer, smart phone, or eBook.

Few people know the definitions of large print or that the idea to include people with impairments in our feasts of worship and the Word comes first from God, from the Bible, not from any secular governments. When our churches are accessible, we maintain our relevance and credibility in our community and in the world. We may even lead the nations.

4) Computer projection: Microsoft (2014) suggests a one-inch projected letter (72 point font on the computer screen) is readable at a distance of 10 feet, a two-inch letter at 20 feet, and a three-inch letter at 30 feet. Expanding line, word, and character spacing helps too. The readability of projected text is affected by the dimensions of a room, the distance between the projector and screen, and a projector's power. Each room may have unique seating, screen size, or ambient lighting issues. A more powerful projector is not always the best solution.

- Creative layout is fun, but when the purpose is to communicate and have people engage the projected or LCD message, make it as readable as possible. Aim for 95% of the population, not just 68%.
- In addition to clearer, universal design for projected slides or on a Liquid Crystal Display (LCD), also make some large print paper copies available.

Slides:

- MediaShout or ProPresenter and other programs are like Microsoft PowerPoint or Corel Presentations in that these programs all use text boxes on a digital slide.
- Use a slide with only one text box and expand the box to fill the slide for a screen or an LCD.
- Use only about 15 words per slide, with a slight margin all around, for song lyrics, sermon notes, scriptures, an outline, or a message. Instead of 50 words on a slide or a 10 point outline on one slide, enlarge the font and use three or more slides, each with about 15 words.
- Keep It Simple Saint.
- Some slides are pictures of text (.jpeg), not real text. The picture can be enlarged to fill the entire slide or the slide can be reformatted as a text slide in order to enlarge the font size. Take a few minutes to rekey it and save a song or outline for improved readability.
- Text from files or an Internet service that automatically go into a list of slides may be too small and need to be rekeyed into a larger font. This is not difficult.

Text:

- Instead of using a font size such as 72, 144, or 216 point font (1, 2, or 3 inches) as standard, fill the entire screen not just part of it, with about 15 words per slide.
- Slightly more or fewer words (12 to 18) per screen can be used so that the portion on each slide makes sense. This also applies to TV, video, or DVDs that include text and will help both seeing and reading.
- Use the font tool to change the font to a larger size and choose a clear, sans serif font type, such as Arial. Do not use fancy fonts or appearance affects such as all caps, italics, outline, shadow, or bold.

- Some people like bold text, but usually the spacing must be expanded, otherwise it can seem that the letters, words, and lines blend into each other.
- Text is harder to read when a slide has pictures, designs, a music staff, multiple colors, video, any movement, or many different font types, or font sizes per slide, or in the same sequence of slides.
- If the text already exists, highlight the text and change the font size and type. It may automatically change, or copy and paste the text with the larger, clearer font onto a new slide or slides.

Color contrast:

- Good color contrast is essential. Yellow or white letters on a plain, medium blue background is preferred by many, but Kitchel (2008) recommends dark letters on a light pastel background. (See this book's cover for an approximate, but inexact example.)
- A color contrast wheel shows the strongest color contrast choices.
- Color laser pointers are invisible to people who are color blind (most often men). Read out loud (ROL).
- Do not use white on black or the reverse which creates a glare that feels like staring into a headlight.

Other considerations:

- The purpose is not to simplify, dumb down, or limit the message. The purpose is to project on each slide a balanced chunk of words that conveys meaning, maybe a whole sentence, or a phrase, but no more than three or four lines, with a font clear and large enough to fill the screen, and with a slight margin all around.
- Do not wear yourself out changing all slides at once; just one, or a few at a time is a good way to begin.

- Consistently make known in an accessible way that paper copies of projected material are available and where they are regularly located. People are unlikely to ask for these or other accommodations.

5) Internet for Computers and Mobile devices: Internet sites should be accessible to people who do not see or hear well, or who have cognitive or ambulatory limitations. Websites may be used without sound, or color, or in high contrast mode, or with the screen magnified, or with a screen reader (no visible screen), or without a mouse or other focus indicator (voice control only), or a combination of these. The site or apps may be used with accessibility features for Microsoft or Apple, or Android, IOS, or Windows smart phones, such as Talkback, or Voice Over. They may be used with Zoomtext, JAWS, a Braille Display or Notetaker.

- Use W3C Web Content Accessibility Guidelines 2.0, AA (WCAG) (see www.w3.org/WAI/intro/components.php)
- Look at accessible Web sites, such as, the American Foundation for the Blind (afb.org) or lighthouseguild.org, or the National Federation of the Blind (nfb.org), or the American Council of the Blind (acb.org), or look at joniandfriends.org, or www.churchinwales.org.uk.
- For more information visit the Google accessibility site, www.google.com/accessibility/ and the Apple accessibility site, http://developer.apple.com/accessibility/.
- See dequeuniversity.com in the USA or usabilitygeek.com in Europe.
- Provide alternative text (alt tags) for images, video, and sound, but not a "Text Only" site which too often is not kept up to date with the main site.
- A link needs a label and please do not use pop-ups.

- Do not use tables for the purpose of layout.
- Design first for mobile devices, tablets, smart phones, and for apps and what is posted with them. The book <u>Mobile First</u> by Luke Wroblewski (2011), explains why it is easier to build small and go larger.
- Test the site initially and as it grows with automated Internet evaluation software, such as, Fireeyes from deque.com. These software programs are not 100% perfect so individuals with disabilities who use various assistive technologies need to periodically test the site.
- Other automated testing services available online are, WAVE – Web Accessibility Versatile Evaluator, by WebAim, Web Accessibility Checker, AChecker - Accessibility Checker, and Amaze by deque.com.
- Make sure the elements of a Web site remain visible, usable, and do not overlap when the screen is enlarged (zoomed) by pressing the "ctrl" and "+" keys together.
- In addition to allowing resizing of text, allow the text to "wrap" or "reflow" without scaling down so that pressing the "ctrl" and "+" keys together will enlarge the text as if it were in a word processor or reflowable eBook, without extending the text beyond the screen sides. This will minimize tedious and tiring sideways scrolling.
- Set up .pdf text so it can be copied, enlarged, and read in a word processor, .pdf documents do not reflow when "zoomed in," or enlarged. They overlap the screen sides and require tedious, tiresome, sideways scrolling.
- If pictures are only being used for visual aesthetics or "eye candy," and contain no content or directional information, include a "skip to content" option.
- Use good semantics. The meaning of headings, titles, and labels is crucial to someone who is not using the picture to understand the message.

- Use more than color or contrast as information, or direction indicators, or for navigation.
- Three screens with about 15 words each, with two or three lines per screen, are easier to read than one screen with 45 words, or five or more lines.
- Include captions and transcriptions with video and audio.
- Set the browser's e-mail default font to a 12 point, sans serif font, such as Arial, without compression.
- A church's Website, Facebook page, or other social media should be accessible. Text is hard to read with small print on top of pictures or textured wallpaper.

6) Hearing Assistance: A hearing Loop, an FM system, and an Infrared system reduce background noise in a hearing aid, making it easier for people to hear clearly, and allow lowering the overall volume to a comfortable level for everyone.

- The cost for any of these three hearing systems is between $1,000 and $1,500. Additional personal receivers as needed for those who do not have hearing aids range from $50 to $100 each or more.
- Installing the wire for a Hearing Loop is not hard, but it is easier in new construction. It may cost $1,000 or more if a contractor is hired to retrofit an auditorium.
- One amplifier for a hearing loop covers a room or an auditorium with a parameter up to 350 feet (107 meters) or 87 feet (27 meters) on each of four walls.
- The cost, $2,000 for a hearing assistance system, can be added to the budget for a sound board, synthesizer, piano, remodeling project, or the cost of a missions trip.
- About half the people in America who need a hearing aid cannot afford one and not all hearing aids have a telecoil that can receive from a hearing loop.
- An FM system for use with up to four headsets may cost $800 for one unit that connects to a sound system.

- Post signs to indicate what system is available and check that it is turned on and works.
- Instead of, or in addition to a hearing system, some churches lend individual assistive listening devices (ALDs) for each church service that cost $50 to $100 each. They help those without a hearing aid, but without a hearing system they amplify all background sounds.
- Real-time captioning of audio material is still imperfect, but it may help, especially with a good transcriptionist.
- A Telecommunication Device for the Deaf (TDD) or a Text Telephone (TTY) for a church office may cost $300.
- The American Sign Language Interpreter Network (www.aslnetwork.com/services/video-relay-interpreting) offers video relay sign language interpreting using high speed internet, a computer, and a USB video camera.
- Lower the key (pitch). Transpose a song from the key of D, to D flat, or C, or B, etc., to make it easier to hear.
- Songs can be transposed with a synthesizer transpose button or a guitar capo. Internet services and computer programs transpose sheet music. Singers and speakers need to clearly articulate the consonants of words.

The headings and 80+ points in these seven bulleted lists can be the basis for an ICT accessibility audit. Also audit the accessibility of activities and architecture. Audit people's knowledge about and willingness to be an accessible church. Use audit checklists that refer to specific behaviors and terms related to the environment and activities of a church. All this ICT accessibility may be too much to apply all at once. It does not cover all impairments such as architecture accessibility for people with mobility impairments (see www.access-board.gov), which in older buildings can be a challenge. It also does not cover access for other special activities (which might be found on www.ada.gov or by contacting specific advocacy groups.)

7) Tactile Writing: The least expensive Braille Embossers cost between, $2,000 and $4,000 and include the software that converts MS Word into Braille. Some software and embossers can print Moon tactile letters as well as Braille.
See www.afb.org/prodBrowseCatResults.asp?CatID=45
for a (2014) discussion and list of Braille Embossers.

- Interpoint Braille uses both sides of the paper.
- Interlinear Braille prints Braille dots and inked text so a person with sight can read the text too.
- The special paper may cost $24 for 500 sheets.
- Braille books require more shelf space.
- Moon writing is not widely used. For more information see http://www.moonliteracy.org.uk/.
- Instead of buying and learning to use a Braille embosser, some churches choose to e-mail their printed material to a Braille or other tactile print service provider and receive the material via the post office.
- Most large cities have multiple sources to contact for Braille transcription.
- Key concerns when outsourcing the work of Brailling are timely, reliable, accurate service.
- Contact www.optasiaministry.org to request a library of Bibles and Bible reference works on a DVD for use with computer screen readers or Braille notetakers.

Less technology may be better for the quality of sound and visuals and for the group experience. Leave the stage. Smaller groups do not have to act like TV or mega churches. Consider teaching a lesson or song antiphonally, that is, responsively speaking or singing, with the congregation repeating a line after the leader. This method is often used to impart our oral tradition and when there is a power outage!

Turning Barriers Into Bridges

Whatever the style or size of a church and whoever may be the intended audience, we can learn to use our computer printing, projection, smart phones, smart boards, e-mail, social media, Websites, and sound systems in inclusive ways. Disability discrimination is illegal where we live. We represent the one true God who cares about those who do not see, hear, or read well.

Unnecessarily excluding people from church is harming them. Misuses of technology may also cause physical harm. A decibel (db) meter, or Sound Pressure Level (SPL) meter, able to measure 30 db to 130 db, can cost $10 to $60. There are SPL apps for smart phones. The accuracy of different sound meters may vary by as much as 20 db. Exposure to 100 db of sound should be avoided to avoid damage to hearing.

Approximate font sizes   John 3:16 (NKJV)
(This is Arial 12 point font) For God so loved the world,
(14 point) that He gave His only begotten Son,
(16 point)   that whoever believes in Him
(18 point)   should not perish
(24)   but have
(36)everlasting life.

---

11. Conclusion

Jesus asked the leaders of a synagogue if it were lawful to do good or to do harm on the Sabbath. They were silent. Their standard was not whether an activity helped someone with a disability. What mattered to them was what activities the law and their traditions permitted on a Sabbath day for those who worshiped God. Jesus was grieved and angered at their hardness of heart (Mark 3:1-6). He pointed to the standard of loving God and their neighbor, but they rejected it and him.

The presence and perspective of people with disabilities exist in scripture. The topic and the people warrant inclusion in any size or style of church. They are part of life. Our hope of resurrection is based on the cross and there is no crucifixion without disability. We cannot fulfill the Great Commandments or the Great Commission without people with disabilities. We do not have religious freedom as a church or as a nation if we block and deny access to worship to millions of our citizens.

Religious institutions benefit from God's gifts of government and law. The churches' environment and context include Biblically based laws restricting how any information and communication technology is used. Such regulation is normal and beneficial. Disability discrimination is a crime in many countries. Nevertheless, the laws of any country are only as good as their enforcement and the moral character of its citizens. We, the Church are called to be the light of the world. Intentionally being an accessible church means we pro-actively remove barriers, implement universal design, and also make reasonable accommodations and adjustments available before anyone requests them and that we talk about this because they are central to who Jesus is and to the Gospel we believe.

Turning Barriers Into Bridges

We cannot assume who is, or who will become a Christian, or who wants or needs to participate in our church gatherings. Wherever a church needlessly disables, hinders, and harms people by blocking their participation she can repent, break the silence, and correct the misuse of technology. After becoming accessible there will still be a need for good works by individuals and by groups. A charity and benevolence ministry will continue to be necessary parts of church life. Various forms of competition can still exist, but we will better represent God and the good news if we create accessible church ministry activities, spaces, and literature, along with evangelistic and missionary outreach, by the inclusive use of information and communication technology.

The Gospel, the good news extended to all, includes the message that a relationship with God is possible for a greater range of people. That includes those whom our Lord Jesus Christ highlighted by words and deeds and specifically told us to bring to our feasts of worship and the Word. Jesus made a way for people who could not see, hear, talk, walk, or think well and so can we. Prepare the way for the people (Isaiah 57:14).

The abundant life Jesus offers us (Jn. 10:10) and the possibilities of a high quality of life with a severe disability are taught, in part, by the church's example. The messages we communicate are important now and for eternity. We can counter a culture of death by creating a community of life. Let us teach a greater understanding of human variation in our classrooms, pulpits, and Christian media and become inclusive by demolishing man-made barriers. We may have failed trying to lead in other areas because of neglecting this one central area. With the Word and the Holy Spirit as our guide, we can learn to do this and lead the nations. We are a church.

More than 300 years ago Isaac Watts wrote, "See how we trifle here below, fond of these earthly toys: our souls, how heavily they go to reach eternal joys." Eternal joy eludes us as we ignore our misuse of information and communication technology. Disability, like death, is a part of life. We fool ourselves if we think silence about it and about disability discrimination are just ways in which we try to reach young people. We are offering them a Christianity without Christ; without him who helped people see and hear and understand the Word of God; and without him whose feet and hands were disabled for us on the cross. We are called to be witnesses of his resurrection. Our silence and misuse of technology must be challenged for the good of those being hurt and for the sake of those who cause the pain and for all who need our message.

> When I survey the wondrous cross
> On which the Prince of Glory died,
> My richest gain I count but loss,
> And pour contempt on all my pride.
>
> When I survey the Wondrous Cross.
> Written by Isaac Watts, 1707, Public Domain.

The above hymn text is in 18 point Arial font. The 27 words of the hymn verse could be enlarged and put in two slide boxes with 13 and 14 words, but presenting and preserving the meaning and the impact of a text requires more than only counting up to 15 words. Accessibility also requires and allows thoughtful consideration of the content.

> Tell of the cross where they nailed Him,
> Writhing in anguish and pain;
> Tell of the grave where they laid Him,
> Tell how He liveth again.
> Love, in that story so tender,
> Clearer than ever I see;
> Stay, let me weep while you whisper,
> "Love paid the ransom for me."
>
> Tell Me the Story of Jesus. Written by
> Fanny J. Crosby, 1915, Public domain

The 50 words of this hymn verse in 14 point font would be clearer if enlarged on four slides of 13, 13, 11, and 13 words.

> Amazing grace how sweet the sound
> that saved a wretch like me!
> I once was lost but now am found,
> was blind but now I see.
>
> Amazing Grace.
> Written by John Newton, 1779, Public Domain

These 26 words could be enlarged onto two slides with 12 and 14 words. We can set up ICT so that more people will find it easier to see, read, and hear new songs or old hymns, message outlines, scriptures, bulletins, and announcements. We can print, project, use sound systems, and the Internet accessibly and be inclusive–and **turn barriers into bridges**.

12. References

Action On Hearing Loss. (2014). Hearing Loss. Retrieved January 11, 2015 from: www.actiononhearingloss.org.uk/.

Albrecht, G., & Devlieger, P. (1999). The Disability Paradox: High quality of life against all odds. Social Science and Medicine, 48(8), 977-988.

American Foundation for the Blind (AFB). (2012). Summary, National Center for Health Statistics for U.S. adults, National Health Interview Survey (NHIS), 2012. Retrieved July 24, 2014 from: www.afb.org.

Americans With Disabilities Act of 1990, As Amended. (2009). (ADAAA, 2008) Retrieved August 11, 2014 from: www.ada.gov/pubs/adastatute08.htm.

Accessibility for Ontarians with Disabilities Act Alliance [AODA Alliance]. (2015). Analysis Of Premier Kathleen Wynne's December 23, 2014 Letter and Economic Development Minister Brad Duguid's December 8, 2014 Letter to the AODA Alliance. Retrieved, January 29, 2015 from: www.aodaalliance.org/strong-effective-aoda.

Anderson, M. [the ICT evangelist] (2013). Perfect ICT every lesson. Independent Thinking Press, an imprint of Crown Publishing, Wales, UK. www.Independentthinkingpress.com.

Barnes, C. (1996). Disability and the myth of the independent researcher. Disability and Society, 11(1), 107-110.

Bagenstos, S. (2004). Has the Americas With Disabilities Act reduced employment for people with disabilities? Berkeley Journal of Employment & Labor Law, 25, 527-563.

Bickenbach, J. (2000). The ADA v. the Canadian Charter of Rights: Disability rights and the Social Model of disability. In L., Francis, & A. Silvers (Eds), (2000) Americans with disabilities: Exploring implications of the law for individuals and institutions. New York: Routledge.

Blanck, P. (2014). eQuality: The struggle for web accessibility by persons with cognitive disabilities. Behavioral Sciences & the Law, 32(1), 4–32.

Brault, M. (2012). Americans With Disabilities: 2010, Current Population Reports, P70-131, U.S. Census Bureau, Washington, DC. (2012 Estimates based on the Survey of Income and Program Participation [SIPP]).

Calvin, J. (1846, 1997). Institutes of the Christian religion. Translation of: Institutio Christianae religionis.; Reprint, with new introduction. Originally published: Edinburgh: Calvin Translation Society, 1845-1846. (III, x, 3). Bellingham, WA: Logos Research Systems, Inc.

Canadian Association of the Deaf. (2012). Statistics On Deaf Canadians. Retrieved January 12, 2015 from: www.cad.ca/statistics_on_deaf_canadians.php.

Canadian Hearing Society (CHS). (2013). Prevalence of Hearing Loss. Retrieved January 15, 2015 from: http://www.chs.ca/facts-and-figures.

Canadian Literacy Network [CLN]. (2015). Literacy Statistics (from IALSS 2003). Retrieved January 12, 2015 from: www.literacy.ca/literacy/literacy-sub/.

References

Charlton, J. (2000). Nothing About Us Without Us: Disability Oppression and Empowerment. University of California Press; New Ed edition.

Church attendance estimates. (2004-13). Weekly statistics. Retrieved: https://en.wikipedia.org/wiki/Church_attendance

Davie. A. (1997, 2016). That All May Worship, An Interfaith Welcome to People with Disabilities. Ginny Thornburgh (Ed.). National Organization On Disability (NOD), and The American Association of People with Disabilities (AAPD), Interfaith Disability Advocacy Coalition: Washington DC. 7th edition (2016), Retrieved May, 2016, from, http://www.aapd.com/wp-content/uploads/2016/03/That-All-May-Worship.pdf.

Dickens, L. (2008). Regulating for equality in employment – is Britain on the right road with the Single Equality Bill? Paper presented at the Third International Conference on Interdisciplinary Social Sciences, Prato, Italy. Retrieved Jan. 29, 2015 from:
www2.warwick.ac.uk/fac/soc/wbs/research/irru/publications/r ecentconf/ld_-_prato_regulating_for_equality_in_employmen t.pdf.

Dickens L. (2014). The Coalition government's reforms to employment tribunals and statutory employment rights —echoes of the past. Retrieved Jan. 30, 2015 from: http://onlinelibrary.wiley.com/doi/10.1111/irj.2014.45.issue-3/ issuetoc.

Edersheim, A. (1896, 2003). The Life and Times of Jesus the Messiah. Longmans, Green and Co., New York, London, and Bombay. Vol. 2, Chapter 15, pgs. 234-237. Bellingham, WA, Logos Research Systems, Inc.

Employment and Social Development Canada. (2015). Indicators of Well-being in Canada. Learning - Adult Literacy. First Results from the Programme for the International Assessment of Adult Competencies (PIAAC), Table B.4.1 Literacy and numeracy - Averages and proficiency levels of population aged 16 to 65 in ALL and PIAAC, Canada, 2003 and 2012, Catalogue no. 89-555-X, Ottawa, 2013.

Employment and Social Development Canada. (2015). Indicators of Well-being in Canada. Canadians in Context - People with Disabilities. Canadian Survey on Disability (CSD, 2012). Retrieved January 10, 2015 from: www.hrsdc.gc. cawww4.rhdcc.gc.ca/indicator.jsp?&indicatorid=40.

Feinberg, J., & Feinberg, P. (1993). Ethics for a Brave New World. Crossway Books, Wheaton IL. Bellingham, WA, Logos Research Systems, Inc.

Frank, J. (1999). Internet or nyet? Some internet prospects and problems in rehabilitation education. *Rehabilitation Education,* 13(2), 163-172.

Frank, J. (2000). Requests for large print accommodations by persons who are visually impaired. Journal of Visual Impairment and Blindness, 94, 716-719.

Frank, J. (2002). What are we teaching about the ADA and what is missing? Re:view, 33, 171-181.

Frank, J. (2006). A survey of the Americans With Disabilities Act (ADA) accommodation request experience of persons who are blind or who have a severe visual impairment. Mississippi State University, MS: Rehabilitation Research and Training Center on Blindness and Low Vision (RRTC-BLV). (now called the National RTC-BLV.)

References

Frank, J., & Bellini, J. (2005). Barriers to the accommodation request process of the Americans With Disabilities Act. Journal of Rehabilitation, 71(2), 28-39. Or: www.questia.com/read/1G1-133317582/barriers-to-the-accommodation-request-process-of-the.

Frank, J., & Stephenson, M. (2013). Let's end disability discrimination in church. The Banner, 148(10), 36-37. Or: www.thebanner.org/features/2013/09/let-s-end-disability-discrimination-in-church.

Gallup Poll (2010). Church Attendance. Retrieved Sept. 4, 2014 from: www.gallup.com/poll/166613/four-report-attending-church-last-week.aspx.

The Hearing Loss Association of America (HLAA). (2014). Who We Are. Communication Access Technology Survey. Retrieved Sept., 2014 from: www.hearingloss.org/sites/default/files/docs/HLAA_CommAccessTech_Survey_Results.pdf.

Indicators of Well-being in Canada. (2012). Learning - Educational Attainment. Retrieved, March 28, 2016 from: http://well-being.esdc.gc.ca/misme-iowb/.3ndic.1t.4r@-eng.jsp?iid=29.

Johnson, M. (2003). Make Them Go Away: Clint Eastwood, Christopher Reeve & The Case Against Disability Rights, Advocado Press Inc.

Johnson, M., & Frank, J. (2004). Stories of accommodation gone wrong. The Ragged Edge, Advocado Press. Retrieved August 5, 2014, from: www.raggededgemagazine.com/ADAaccores1004.html.

Kagle, J., & Cowger, C. (1984). Blaming the client: Implicit agenda in practice and research? Social Work, 29,347-351.

Kaye, S. (1998). Is the Status of People with Disabilities Improving? Disabilities Statistics Center, University of California, San Francisco Abstract 21. Retrieved August 13, 2014 from: http://dsc.ucsf.edu/publication.php?pub_id=7.

Kistemaker, S. (1986, 2001). Vol. 14: New Testament commentary: Exposition of James and the Epistles of John. New Testament Commentary (260). Grand Rapids: Baker Book House. Bellingham, WA, Logos Research Systems, Inc.

Kitchel, J. (2004). APH guidelines for print document design. American Printing House for the Blind. Retrieved August 11, 2014 from: www.aph.org/edresearch/lpguide.htm.

Kitchel, J. (2008). Color and text guidelines for the development of Power Point, computer and Web page presentations and text applications for audiences that may include persons with low vision. The American Printing House for the Blind, Louisville, KY. Retrieved August 11, 2014 from: www.aph.org/edresearch/ppt_guidelines.ppt.

Koster, S. (2005). Calvin Institute of Christian Worship, 2003, Projectors in worship: Survey summary: Results of research on the use of technology in worship. June 25, 2005. Retrieved, July 22, 2014 from: http://worship.calvin.edu/resources/resource-library/projectors-in-worship-survey-summary/.

Matson, F. (1990). Walking Alone and Marching Together: A History of the Organized Blind Movement in the United States, 1940-1990. National Federation of the Blind.

Microsoft. (2014). Tips for creating and delivering an effective presentation - PowerPoint. Retrieved August 11, 2014, from: http://office.microsoft.com/en-us/powerpoint-help/tips-for-creating-and-delivering-an-effective-presentation-HA010207864.aspx.

References

Moss, K., Burris, S. Ullman, M., Johnsen, M., & Swanson, J. (2001). Unfunded mandate: An empirical study of the implementation of the Americans with Disabilities Act by the Equal Employment Opportunity Commission. Kansas Law Review, November, 50(1). 1-110.

National Association of Evangelicals (NAE). (2015). Code of Ethics for Congregations and Their Leadership Teams. retrieved Oct. 7, 2015 from: http://nae.net/code-of-ethics-for-congregations/

National Health Statistics [NHS]. (2013). Vision impairment. Retrieved Jan 12, 2015 from: www.nhs.uk/conditions/ Visual-impairment/Pages/Introduction.aspx.

National Highway Traffic Safety Administration. (2014). Fatality Analysis Reporting System Retrieved August 29, 2014 from: www-fars.nhtsa.dot.gov/Main/index.aspx.

National Institute on Deafness and Other Communication Disorders. (2014). Quick Statistics. Retrieved December 12, 2014, from; www.nidcd.nih.gov/health/statistics/pages/quick.aspx.

National Institute of Literacy. (2013). Adult Literacy Survey. United States Department of Education. Retrieved July 24, 2014 from: www.statisticbrain.com/number-of-american-adults-who-cant-read/.

National Literacy Trust. (2011). Literacy in Britain. Retrieved Jan. 12, 2015 from:www.literacytrust.org.uk/adult_literacy/ illiterate_adults_in_england.

O'Brien, R. (2001). Crippled Justice: The history of modern disability policy. Chicago: The University of Chicago Press.

O'Connor, C. (2013). Where is the accessible, tactile currency? Retrieved July 22 2014, from; www.pdrib.com/blog/where-is-the-accessible-tactile-currency/.

Office of National Statistics [ONS]. (2012). Statistical bulletin: Adult Health in Great Britain 2012. Retrieved Jan 12, 2015 from: www.ons.gov.uk/ons/rel/ghs/opinions-and-lifestyle-survey/adult-health-in-great-britain--2012/stb-health-2012.html.

Potok, A. (2002). A matter of dignity. New York: Bantam Books/Random House.

Rogers, E. (2003). Diffusion of Innovations (5th ed.) Simon and Schuster.

Rubin, S., & Roessler, R. (2001). Foundations of the Vocational Rehabilitation Process. (5th ed.) Austin, TX: pro-ed. Chapter 2 & chapter 4.

Ryan, W. (1971). Blaming the Victim. New York: Pantheon Books.

Sanchez, W., & Fried, J. (1997). Giving voice to students' narratives. College Teaching, 45 (1), 26.

Shapiro, J. (1994). No Pity: People with Disabilities Forging a New Civil Rights Movement. Broadway books.

Singleton, P. (2009). Insult to Injury: Disability, earnings, and divorce. Syracuse University, Department of Economics. Retrieved July 22, 2014, from: http://papers.ssrn.com/sol3/papers.cfm?abstract_id=1553246.

Stade, J. (2014). Eye Health Data Summary: A review of published data in England. Royal National Institute of Blind People [RNIB]. Retrieved January 12, 2015 from: www.rnib.org.uk/sites/default/files/Eye_health_data _summary_report_ 2014.pdf.

Statistics Canada. (2014). Canada's population estimates: Age 2014. Retrieved January 12, 2015, from: www.statcan.gc.ca/ daily-quotidien/140926/dq140926b-eng.htm.

Statistics Canada. (2014). Population by year by province and by territory. Retrieved January 12, 2015 from: www.statcan.gc.ca/tables-tableaux/sum-som/l01/cst01/demo 02a-eng.htm.

Statistics Canada. (2010). Participation and Activity Limitation Survey (PALS) 2006, Retrieved Jan 10, 2015 at ww.statcan.gc.ca/pub/89-628-x/89-628-x2010015-eng.htm.

Stumbo, E. (2014). Confessions of a pastor's wife: The church is forgetting us. Retrieved July 25, 2014 from: www.ellenstumbo.com/confessions-pastors-wife-church-forgetting-us/.

Temple, G., with Ball, L. (2012). Enabling Church. www.torchtrust.org, or the Society for promoting Christian Knowledge, www.spckpublishing.co.uk.

United States Access Board. (2002). ADA Accessibility Guidelines (ADAAG), section 4.3 Signage. Retrieved August 14, 2014 from: www.access-board.gov/guidelines-and-standards/buildings-and-sites/about-the-ada-standards/background/adaag#4.30s.

United States Department of Justice. (2012). 2010 ADA Standards for Accessible Design, Retrieved August 11, 2014 from: www.ada.gov/2010ADAstandards_index.htm.

Vander Plaats, D. (2014). 5 Stages: The journey of disability attitudes. Elim Christian Services. Retrieved August 13, 2014 from: www.elimcs.org/news/elim-worship-team-chicago-sun-times.

Wright, B. (1981). Value laden beliefs and principles for rehabilitation. In S. Regnier & M. Petkovsek (comp.) (1985), 25 years of concepts, principles, precepts. A collection of articles published in rehabilitation literature 1959-1984. (pp 113-116). Chicago, IL: National Easter Seals Society.

Yong, A. (2010). Disability, the Bible and the Church. Eerdmans, Grand Rapids MI, pages. 141-142.

"I long to accomplish a great and noble task, but it is my chief duty to accomplish small tasks as if they were great and noble." – Helen Keller.

"There are many of us that are willing to do great things for the Lord, but few of us are willing to do little things." – Dwight L. Moody.

"If revival is being withheld from us, it is because some idol remains still enthroned; because we still insist in placing our reliance in human schemes; because we still refuse to face the unchangeable truth that, 'It is not by might, but by My Spirit." – Jonathan Goforth.

"Someone asked, Will the heathen who have never heard the Gospel be saved? It is more a question with me whether we – who have the Gospel and fail to give it to those who have not – can be saved." – Charles Spurgeon.

About the Author: John Jay Frank, an Ordained Minister of the Gospel and a Certified Rehabilitation Counselor for many years, has had training and experience in Reading Education, Biblical Studies, Information and Communication Technology, and Music. He earned a Ph.D. in Rehabilitation Counseling at Syracuse University in New York (2003). Beyond experience with his own vision impairment, his professional and volunteer work includes the direct care of people with various severe impairments, as well as pastorate, chaplaincy, research and teaching positions, and missionary and evangelistic activities.

Dr. Frank is the founder of Minstrel Missions LLC. He writes articles and books, and records songs and hymns for the education, edification, encouragement, and comfort of the body of Christ. See: www.minstrelmissions.com, or e-mail comments and questions to: minstrelmissions@gmail.com

3 Books:

Turning Barriers Into Bridges: The Inclusive Use
    of Information and Communication Technology
    for Churches in America, Britain, and Canada

Myths, Lies, & Denial:
    Christian and Secular Counseling in America

A Minstrel's Notes:
    Stories and Sermons on Worship in Spirit and In Truth,
    and Music Theory and Technique for the Acoustic Guitar

6 Music Recordings (CDs):

For All God's Children	Hymns of the Church
Comfort Ye My People	Hymns and Carols
The Cross Of Life	Hymns of God's Grace

Love one another as I have loved you
(John 15:12).

Go therefore and make disciples
(Matt. 28:18-20).

Made in the USA
Charleston, SC
09 December 2016